PRACTICAL GENETIC ALGORITHMS

PRACTICAL GENETIC ALGORITHMS

RANDY L. HAUPT and SUE ELLEN HAUPT

A Wiley-Interscience Publication

JOHN WILEY & SONS, INC.

New York / Chichester / Weinheim / Brisbane / Singapore / Toronto

Library of Congress Cataloging-in-Publication Data:

Haupt, Randy L.
    Practical genetic algorithms / Randy L. Haupt & Sue Ellen Haupt.
      p. cm.
    "A Wiley-Interscience publication."
    Includes bibliographical references and index.
    ISBN 0-471-18873-5 (cloth : alk. paper)
    1. Genetic algorithms.   I. Haupt, S. E.   II. Title.
  QA402.5.H387   1998
  519.7—dc21   97-13172
                                             CIP

Printed in the United States of America

10 9 8 7 6 5 4 3 2

To our parents,
Anna Mae and Howard Haupt
Iona and Charles Slagle
and
Our offspring,
Bonny Ann and Amy Jean Haupt

CONTENTS

PREFACE

The book has been organized to take the genetic algorithm in stages. Chapter 1 lays the foundation for the genetic algorithm by discussing numerical optimization and introducing some of the traditional minimum seeking algorithms. Next, the idea of modeling natural processes on the computer is introduced through a discussion of annealing and the genetic algorithm. A brief genetics background is supplied to help the reader understand the terminology and rationale for the genetic operators. The genetic algorithm comes in two flavors: binary parameter and real parameter. Chapter 2 provides an introduction to the binary genetic algorithm, which is the most common form of the algorithm. Parameters are quantized, so there are a finite number of combinations. This form of the algorithm is ideal for dealing with parameters that can assume only a finite number of values. Chapter 3 introduces the continuous parameter genetic algorithm. This algorithm allows the parameters to assume any real value within certain constraints. Chapter 4 uses the algorithms developed in the previous chapters to solve some problems of interest to engineers and scientists. Chapter 5 returns to building a good genetic algorithm, extending and expanding upon some of the components of the genetic algorithm. Chapter 6 attacks more difficult technical problems. Finally, Chapter 7 surveys some of the current extensions to genetic algorithms and applications, and gives advice on where to get more information on genetic algorithms. Some aids are supplied to further help the budding genetic algorithmist. Appendix A lists some genetic algorithm routines in pseudocode. A glossary and a list of symbols used in this book are also included.

We are indebted to several friends and colleagues for their help. First, our thanks goes to Dr. Christopher McCormack of Rome Laboratory for introducing us to genetic algorithms several years ago. The idea for writing this book and the encouragement to write it, we owe to Professor Jianming Jin of the University of Illinois. Finally, the excellent reviews by Professor Daniel Pack, Major Cameron Wright, and Captain Gregory Toussaint of the United States Air Force Academy were invaluable in the writing of this manuscript.

RANDY L. HAUPT
SUE ELLEN HAUPT

Reno, Nevada
September 1997

LIST OF SYMBOLS

$chromosome_n$	Vector containing the parameters
$cost$	Cost associated with a parameter set
$cost_{min}$	Minimum cost of a chromosome in the population
$cost_{max}$	Maximum cost of a chromosome in the population
c_n	Cost of chromosome n
C_n	Normalized cost
dad	Vector containing row numbers of chromosomes selected as one mate
$f(*)$	Cost function
G	Generation gap
$g_m(x, y, \ldots)$	Constraints on the cost function
$gene[m]$	Binary version of p_n
hi	Highest number in the parameter range
$IPOP$	Initial population matrix
$life_{min}$	Minimum lifetime of a chromosome
$life_{max}$	Maximum lifetime of a chromosome
lo	Lowest number in the parameter range
M_{iter}	Number of generations until the genetic algorithm stops
mom	Vector containing row numbers of chromosomes selected as one mate

N_{bad}	Number of chromosomes removed from the population each generation
N_{bits}	$N_{gene} \times N_{par}$
N_{gene}	Number of bits in the gene
N_{good}	Number of chromosomes in the mating pool
N_{ipop}	Number of chromosomes in the initial population
N_{par}	Number of parameters
N_{pop}	Number of chromosomes in the population from generation to generation
$offspring_n$	Child created from mating two chromosomes
$parent_n$	A parent selected to mate
p_n	Parameter n
p_{norm}	Normalized parameter
p_{quant}	Quantized parameter
p_{lo}	Smallest parameter value
p_{hi}	Highest parameter value
P_n	Probability of chromosome n being selected for mating
P_t	Probability of the schema being selected to survive to the next generation
q_n	Quantized version of p_n
Q_i	Number of different values that parameter i can have
Q_t	Probability of the schema being selected to mate
R_t	Probability of the schema not being destroyed by crossover or mutation
s_t	Number of schemata in generation t
T	Total number of different parameter combinations
X_{rate}	Crossover rate
$\hat{x}, \hat{y}, \hat{z}$	Unit vectors in the x, y, and z directions
α	Parameter where crossover occurs
β	Mixing value for continuous parameter crossover
∇	$\frac{\partial f}{\partial x}\hat{x} + \frac{\partial f}{\partial y}\hat{y} + \frac{\partial f}{\partial z}\hat{z}$
∇^2	$\frac{\partial^2 f}{\partial x^2} + \frac{\partial^2 f}{\partial y^2} + \frac{\partial^2 f}{\partial z^2}$
γ_n	Nonnegative scalar that minimizes the function in the direction of the gradient
λ	Lagrange multiplier
η	$\frac{1}{2}(life_{max} - life_{min})$
μ	Mutation rate

CHAPTER 1

INTRODUCTION TO OPTIMIZATION

Optimization is the process of making something better. An engineer or scientist conjures up a new idea and optimization improves on that idea. Optimization consists of trying variations on an initial concept and using the information gained to improve on the idea. A computer is the perfect tool for optimization as long as the idea or parameter influencing the idea can be input in electronic format. Feed the computer some data and out comes the solution. Is this the only solution? Often times not. Is it the best solution? That's a tough question. Optimization is the math tool that helps us answer these questions.

This chapter begins with an elementary explanation of optimization, then moves on to a historical development of minimum-seeking algorithms. A seemingly simple example reveals many shortfalls of the minimum seekers. Then the natural algorithms, simulated annealing and genetic algorithms, which break from the traditional approach by modeling natural processes and invoking stochastic rules, are introduced. We show how they avoid the problems of the traditional minimum-seeking algorithms. The chapter ends with a brief introduction to biological genetics, the basis of the genetic algorithm.

1.1 FINDING THE BEST SOLUTION

The terminology "best" solution implies that there is more than one solution and the solutions are not of equal value. The definition of best is relative to the problem at hand, its method of solution, and the tolerances allowed. Thus, the optimal solution depends upon the person formulating the problem. Education, opinions, bribes, and amount of sleep are factors influencing the definition of best. Some problems have exact answers or roots, and best has an exact definition. Examples include best home run hitter in baseball and a solution to a linear first-order differential equation. Other problems have various minimum or maximum solutions known as optimal points or extremum, and best may be a relative definition. Examples include best piece of artwork or best baseball player.

1.1.1 What is Optimization?

Our lives confront us with many opportunities for optimization. What time do we get up in the morning so that we maximize the amount of sleep yet still make it to work on time? What is the best route to work? Which project do we tackle first? When designing something, we shorten the length of this or reduce the weight of that, as we want to minimize the cost or maximize the appeal of a product. Optimization is the process of adjusting the inputs to or characteristics of a device, mathematical process, or experiment to find the minimum or maximum output or result (Figure 1.1). The input consists of parameters; the process or function is known as the cost function, objective function, or fitness function; and the output is the cost or fitness. If the process is an experiment, then the parameters are physical inputs to the experiment.

For most of the examples in this book, we define the output from the process or function as the cost. Since cost is something to be minimized, optimization becomes minimization. Sometimes maximizing a function

Figure 1.1 Diagram of a function or process that is to be optimized. Optimization varies the input to achieve a desired output.

makes more sense. To maximize a function, just slap a minus sign on the front of the output and minimize. As an example, maximizing $1 - x^2$ over $-1 \leq x \leq 1$ is the same as minimizing $x^2 - 1$ over the same interval. Consequently, in this book we occasionally address the maximization of some function.

Life is interesting due to the many decisions and seemingly random events that take place. Quantum theory suggests there are an infinite number of dimensions, and each dimension corresponds to a decision made. Life is also highly nonlinear, so chaos plays an important role, too. A small perturbation in the initial condition may result in a very different and unpredictable solution. These theories suggest a high degree of complexity encountered when studying nature or designing products. Science developed simple models to represent certain limited aspects of nature. Most of these simple (and usually linear) models have been optimized. In the future, scientists and engineers must tackle the unsolvable problems of the past, and optimization is the primary tool needed in the intellectual toolbox.

1.1.2 Root Finding vs. Optimization

Approaches to optimization are akin to root or zero finding methods, only harder. For the one-parameter case, only two points are needed to bracket a zero (one negative and the other positive). On the other hand, bracketing an extremum (maximum or minimum) requires three points (the middle point having a lower or higher value than either end point). In the mathematical approach, root finding requires searching for zeros of a function, while optimization requires finding zeros of its derivatives. Computing the derivative is not always an easy task. We like the simplicity of root finding problems, so we teach root finding techniques to students of engineering, math, and science courses. Many technical problems are formulated to find roots, when they might be more naturally posed as optimization problems. However, the toolbox containing the optimization tools is small and inadequate.

Another difficulty with optimization is determining if a given minimum is the best (global) minimum or a suboptimal (local) minimum. Root finding doesn't have this difficulty. One root is as good as another (by definition, all roots cause the function to be zero).

Finding the minimum of a nonlinear function is especially difficult. Typical approaches to highly nonlinear problems involve either linearizing the problem in a very confined region or restricting the optimization to a small region. In short, we cheat.

1.1.3   Categories of Optimization

Figure 1.2 shows that optimization algorithms may be broken into six categories. Neither these six views nor their branches are necessarily mutually exclusive. For instance, a dynamic optimization problem could be either constrained or unconstrained. In addition, some of the parameters may be discrete and others continuous. Let's begin at the top left of Figure 1.2 and work our way around clockwise.

1. Trial-and-error optimization refers to the process of adjusting parameters that affect the output without knowing much about the process that produces the output. An example is adjusting the rabbit ears on a TV to get the best picture and audio reception. An antenna engineer can only guess at why certain contortions of the rabbit ears result in a better picture than other contortions. Experimentalists prefer this approach. Many great discoveries have been made using this trial-and-error approach to optimization, such as the discovery and refinement of penicillin as an antibiotic. In contrast, mathematical function optimization assumes that we can describe a process by a mathematical formula. Various mathematical methods are applied to the function to find the optimum solution. This approach is preferred by theoreticians.

2. If there is only one parameter, the optimization is one-dimensional. A problem having more than one parameter requires multidimensional optimization. Optimization becomes increasingly difficult as the number of dimensions increases. Many multidimensional optimization approaches generalize to a series of one-dimensional approaches.

3. Dynamic optimization means that the output is a function of time, while static means that the output is independent of time. When living in

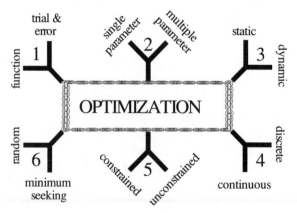

Figure 1.2   Six categories of optimization algorithms.

the suburbs of Boston, there were several ways to drive back and forth to work. What was the best route? From a distance point of view, the problem is static, and the solution can be found using a map or the odometer of a car. In practice, this problem is not simple because of the myriad of variations in the routes. The shortest route isn't necessarily the fastest route, though. Finding the fastest route is a dynamic problem whose solution depends upon the time of day, the weather, accidents, and so on. The static problem is difficult to solve for the best solution, but the added dimension of time increases the challenge of solving the dynamic problem.

4. Optimization can also be distinguished by either discrete or continuous parameters. Discrete parameters have only a finite number of possible values, whereas continuous parameters have an infinite number of possible values. If we are deciding in what order to attack a series of tasks on a list, discrete optimization is employed. Discrete parameter optimization is also known as combinatorial optimization, because the optimum solution consists of a certain combination of parameters from the finite pool of all possible parameters. However, if we are trying to find the minimum value of $f(x)$ on a number line, it is more appropriate to view the problem as continuous.

5. Parameters often have limits or constraints. Constrained optimization incorporates parameter equalities and inequalities into the cost function. Unconstrained optimization allows the parameters to take any value. A constrained parameter often converts into an unconstrained parameter through a transformation of variables. Most numerical optimization routines work best with unconstrained parameters. Consider the simple constrained example of minimizing $f(x)$ over the interval $-1 \leq x \leq 1$. The parameter converts x into an unconstrained parameter u by letting $x = \sin(u)$ and minimizing $f(\sin(u))$ for any value of u. When constrained optimization formulates parameters in terms of linear equations and linear constraints, it is called a linear program. When the cost equations or constraints are nonlinear the problem becomes a nonlinear programming problem.

6. Some algorithms try to minimize the cost by starting from an initial set of parameter values. These minimum seekers easily get stuck in local minima but tend to be fast. They are the traditional optimization algorithms and are generally based on calculus methods. Moving from one parameter set to another is based on some determinant sequence of steps. On the other hand, random methods use some probabilistic calculations to find parameter sets. They tend to be slower but have greater success at finding the global minimum.

1.2 MINIMUM SEEKING ALGORITHMS

Searching the cost surface (all possible function values) for the minimum cost lies at the heart of all optimization routines. Usually, a cost surface has many peaks, valleys, and ridges. An optimization algorithm works much like a hiker trying to find the minimum altitude in Rocky Mountain National Park. Starting at some random location within the park, the goal is to intelligently proceed to find the minimum altitude. There are many ways to hike or glissade to the bottom from a single random point. Once the bottom is found, though, there is no guarantee that an even lower point doesn't lie over the next ridge. Certain constraints, such as cliffs and bears, influence the path of the search as well. Pure downhill approaches usually fail to find the global optimum unless the cost surface is quadratic (bowl-shaped).

There are many good texts which describe optimization methods (for example; Press, et al., 1992; Cuthbert, 1987). A history is given by Boyer and Merzbach (1991). Here we give a very brief review of the development of optimization strategies.

1.2.1 Exhaustive Search

The brute force approach to optimization looks at a sufficiently fine sampling of the cost function to find the global minimum. It is equivalent to spending the time, effort, and resources to thoroughly survey Rocky Mountain National Park. In effect, a topographical map can be generated by connecting lines of equal elevation from an interpolation of the sampled points. This exhaustive search requires an extremely large number of cost function evaluations to find the optimum. For example, consider solving the two-dimensional problem

Find the minimum of: $\quad f(x, y) = x \sin(4x) + 1.1y \sin(2y) \quad$ (1.1)

Subject to: $\quad 0 \le x \le 10 \quad$ and $\quad 0 \le y \le 10 \quad$ (1.2)

Figure 1.3 shows a three-dimensional plot of equation (1.1) in which x and y are sampled at intervals of 0.1, requiring a total of 101^2 function evaluations. This same graph is shown as a contour plot with the global minimum marked by an asterisk in Figure 1.4. In this case, the global minimum is easy to see. Graphs have aesthetic appeal, but are only practical for one- and two-dimensional cost functions. Usually, a list of function values is generated over the sampled parameters, then the list is searched for the minimum value. The exhaustive search does the surveying necessary to

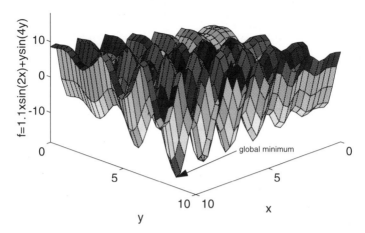

Figure 1.3   Three-dimensional plot of equation (1.1) in which x and y are sampled at intervals of 0.1.

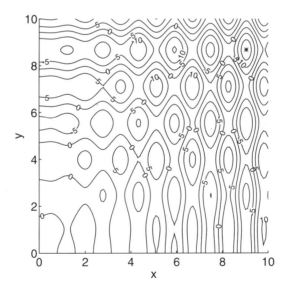

Figure 1.4   Contour plot of equation (1.1).

produce an accurate topographic map. This approach requires checking an extremely large but finite solution space with the number of combinations of different parameter values given by

$$T = \prod_{i=1}^{N_{par}} Q_i \qquad (1.3)$$

where

T = total number of different parameter combinations

N_{par} = number of different parameters

Q_i = number of different values that parameter i can attain

With fine enough sampling, exhaustive searches don't get stuck in local minima and work for either continuous or discontinuous parameters. However, they take an extremely long time to find the global minimum. Another shortfall of this approach is that the global minimum may be missed due to undersampling. It is easy to undersample when the cost function takes a long time to calculate. Hence, exhaustive searches are only practical for a small number of parameters in a limited search space.

A possible refinement to the exhaustive search includes first searching a coarse sampling of the fitness function, then progressively narrowing the search to promising regions with a finer tooth comb. This approach is similar to first examining the terrain from a helicopter view, then the valleys are surveyed but not the peaks and ridges. This approach speeds convergence and increases the number of parameters that can be searched, but at the same time, increases the odds of missing the global minimum. Most optimization algorithms employ a variation of this approach and start exploring a relatively large region of the cost surface (take big steps), then contract the search around the best solutions (take smaller and smaller steps).

1.2.2   Analytical Optimization

Calculus provides the tools and elegance for finding the minimum of many cost functions. Simplifying the thought process to a single parameter for a moment, an extremum is found by setting the first derivative of a cost function to zero and solving for the parameter value. If the second derivative is greater than zero, the extremum is a minimum, and conversely, if the second derivative is less than zero, the extremum is a maximum. One way to find the extrema of a function of two or more parameters is to take the gradient of the function and set it equal to zero ($\nabla f(x, y) = 0$). For example, taking the gradient of equation (1.1) results in

$$\frac{\partial f}{\partial x} = \sin(4x_m) + 4x_m \cos(4x_m) = 0, \qquad 0 \le x \le 10$$

and
(1.4)

$$\frac{\partial f}{\partial y} = 1.1 \sin(2y_m) + 2.2y_m \cos(2y_m) = 0, \qquad 0 \le y \le 10$$

Next, these equations are solved for their roots, x_m and y_m, which are a family of lines. Extrema occur at the intersection of these lines. Note that these transcendental equations may not always be separable, making it very difficult to find the roots. Finally, the Laplacian of the function is calculated.

$$\frac{\partial^2 f}{\partial x^2} = 8 \cos 4x - 16x \sin 4x, \qquad 0 \le x \le 10$$

and
(1.5)

$$\frac{\partial^2 f}{\partial y^2} = 4.4 \cos 2y - 4.4y \sin 2y, \qquad 0 \le y \le 10$$

The roots are minima when $\nabla^2 f(x_m, y_m) > 0$. Unfortunately, this process doesn't give a clue as to which of the minima is a global minimum. Searching the list of minima for the global minimum makes the second step of finding $\nabla^2 f(x_m, y_m)$ redundant. Instead, $f(x_m, y_m)$ is evaluated at all the extrema, then the list of extrema is searched for the global minimum. This approach is mathematically elegant compared to the exhaustive or random searches. It quickly finds a single minimum, but requires a search scheme to find the global minimum. Continuous functions with analytical derivatives are necessary (unless derivatives are taken numerically, which results in even more function evaluations plus a loss of accuracy). If there are too many parameters, then it is difficult to find all the extrema. The gradient of the cost function serves as the compass heading pointing to the steepest downhill path. It works well when the minimum is nearby, but cannot deal well with cliffs or boundaries, where the gradient can't be calculated.

In the eighteenth century, Lagrange introduced a technique for incorporating the equality constraints into the cost function. The method now known as Lagrange multipliers finds the extrema of a function $f(x, y, \dots)$ with constraints $g_m(x, y, \dots) = 0$, by finding the extrema of the new function $F(x, y, \dots, \lambda_1, \lambda_2, \dots) = f(x, y, \dots) + \sum_{m=1}^{M} \lambda_m g_m(x, y, \dots)$ (Borowski and Borwein, 1991). Then, when gradients are taken in terms of the new parameters, λ_m, the constraints are automatically satisfied.

As an example of this technique, consider equation (1.1) with the constraint $x + y = 0$. The constraints are added to the cost function to produce the new cost function

$$f_\lambda = x \sin(4x) + 1.1y \sin(2y) + \lambda (x + y) \tag{1.6}$$

Taking the gradient of this function of three variables yields

$$\frac{\partial f}{\partial x} = \sin(4x_m) + 4x_m \cos(4x_m) + \lambda = 0$$

$$\frac{\partial f}{\partial y} = 1.1 \sin(2y_m) + 2.2y_m \cos(2y_m) + \lambda = 0 \tag{1.7}$$

$$\frac{\partial f}{\partial \lambda} = x_m + y_m = 0$$

These equations combine to

$$4x_m \cos(4x_m) + \sin(4x_m) + 1.1 \sin(2x_m) + 2.2x_m \cos(2x_m) = 0 \tag{1.8}$$

where $(x_m, -x_m)$ are the minima of equation (1.6). The solution is once again a family of lines crossing the domain.

The many disadvantages to the calculus approach make it an unlikely candidate to solve most optimization problems encountered in the real world. Even though it is impractical, most numerical approaches are based on it. Typically, an algorithm starts at some random point in the search space, calculates a gradient, then heads downhill to the bottom. These numerical methods head downhill fast; however, they often find the wrong minimum (a local minimum rather than the global minimum) and don't work well with discrete parameters. Gravity helps us find the downhill direction when hiking, but we will still most likely end up in a local valley in the complex terrain.

Calculus-based methods were the bag of tricks for optimization theory until von Neumann developed the minimax theorem in game theory (Thompson, 1992). Games require an optimum move strategy to guarantee winning. That same thought process forms the basis for more sophisticated optimization techniques. In addition, techniques were needed to find the minimum of cost functions having no analytical gradients. Shortly before and during World War II, Kantorovich, von Neumann, and Leontief solved linear problems in the fields of transportation, game theory, and input-output models (Anderson, 1992). Linear programming concerns the minimization of a linear function of many variables subject to constraints that are linear equations and equalities. In 1947, Dantzig introduced the simplex method, which has been the workhorse for solving linear programming problems (Williams, 1993). This method has been widely implemented in computer codes since the mid-1960s.

Another category of methods is based on integer programming, an extension of linear programming in which some of the parameters can only take integer values (Williams, 1993). Nonlinear techniques were also under investigation during World War II. Karush extended Lagrange multipliers to constraints defined by equalities and inequalities, so a much larger category of problems could be solved. Kuhn and Tucker improved and popularized this technique in 1951 (Pierre, 1992). In the 1950s, Newton's method and the method of steepest descent were commonly used.

1.2.3   Nelder–Mead Downhill Simplex Method

The development of computers spurred a flurry of activity in the 1960s. In 1965 Nelder and Mead introduced the downhill simplex method (Nelder and Mead, 1965) that doesn't require the calculation of derivatives. A simplex is the most elementary geometrical figure that can be formed in dimension n and has $n + 1$ sides (e.g., a triangle in two-dimensional space). Each iteration generates a new vertex for the simplex. If this new point is better than at least one of the existing vertices, it replaces the worst vertex. In this way, the diameter of the simplex gets smaller and the algorithm stops when the diameter reaches a specified tolerance. This algorithm is not known for its speed, but it has a certain robustness that makes it attractive.

Since the Nelder–Mead algorithm gets stuck in local minima, it will be combined with the random search algorithm to find the minimum to equations (1.1) and (1.2). Assuming that there is no prior knowledge of the cost surface, a random first guess is as good as any place to start. How close does this guess have to be to the true minimum before the algorithm can find it? Some simple experimentation helps us arrive at this answer. The first column in Table 1.1 shows twelve random starting values, the ending values, and the final costs. None of the trials arrived at the global minimum.

TABLE 1.1 Comparison of Nelder–Meade and Conjugate Gradient Algorithms*

		Nelder–Mead			BFGS		
Starting Point		Ending Point			Ending Point		
x	y	x	y	cost	x	y	cost
8.9932	3.7830	9.0390	5.5427	−15.1079	9.0390	2.4567	−11.6835
3.4995	8.9932	5.9011	2.4566	−8.5437	5.9011	2.4566	−8.5437
0.4985	7.3803	1.2283	8.6682	−10.7228	0.0000	8.6682	−9.5192
8.4066	5.9238	7.4696	5.5428	−13.5379	9.0390	5.5428	−15.1079
0.8113	6.3148	1.2283	5.5427	−7.2760	0.0000	5.5428	−6.0724
6.8915	1.8475	7.4696	2.4566	−10.1134	7.4696	2.4566	−10.1134
7.3021	9.5406	5.9011	8.6682	−15.4150	7.4696	8.6682	−16.9847
5.6989	8.2893	5.9011	8.6682	−15.4150	5.9011	8.6682	−16.9847
6.3245	3.2649	5.9011	2.4566	−8.5437	5.9011	2.4566	−8.5437
5.6989	4.6725	5.9011	5.5428	−11.9682	5.9011	5.5428	−11.9682
4.0958	0.3226	4.3341	0.0000	−4.3269	4.3341	0.0000	−4.3269
4.2815	8.2111	4.3341	8.6622	−13.8461	4.3341	8.6622	−13.8461
average				−11.2347			−11.1412

*Frequently, both methods find the same minimum. Neither algorithm found the global minimum in 12 runs.

Box (1965) extended the simplex method and called it the complex method, which stands for constrained simplex method. This approach allows the addition of inequality constraints, uses up to $2n$ vertices, and expands the polyhedron at each normal reflection.

1.2.4 Optimization Based on Line Minimization

The largest category of optimization methods fall under the general title of successive line minimization methods. An algorithm begins at some random point on the cost surface, chooses a direction to move, then moves in that direction until the cost function begins to increase. Next, the procedure is repeated in another direction. Devising a sensible direction to move is critical to algorithm convergence and has spawned a variety of approaches.

A very simple approach to line minimization is the coordinate search method (Schwefel, 1995). It starts at an arbitrary point on the cost surface, then does a line minimization along the axis of one of the parameters. Next, it selects another parameter and does another line minimization along that axis. This process continues until a line minimization is done along each of the parameters. Then, the algorithm cycles through the parameters until an acceptable solution is found. Figure 1.5 models a possible path the algorithm might take in a quadratic cost surface. In general, this method is slow.

Rosenbrock (1960) developed a method that does not limit search directions to be parallel to the parameter axes. The first iteration of the Rosenbrock method uses coordinate search along each parameter to find the first improved point (see Figure 1.6). The coordinate axes are then rotated until the first new coordinate axis points from the starting location to the first point. Gram-Schmidt orthogonalization finds the directions of the other new coordinate axes based on the first new coordinate axis. A coordinate search is then performed along each new coordinate axis. As before, this process iterates until an acceptable solution is found.

The steepest descent algorithm starts at an arbitrary point on the cost surface and minimizes along the direction of the gradient. This approach has been extremely popular and originated with Cauchy in 1847. The simple formula for the $(n + 1)$th iteration is given by

$$\begin{bmatrix} x_{n+1} \\ y_{n+1} \end{bmatrix} = \begin{bmatrix} x_n \\ y_n \end{bmatrix} - \gamma_n \nabla f(x_n, y_n) \qquad (1.9)$$

where γ_n is a nonnegative scalar that minimizes the function in the direction of the gradient. By definition, the new gradient formed at each iteration

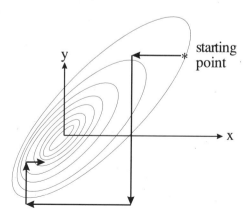

Figure 1.5 Possible path that the coordinate search method might take on a quadratic cost surface.

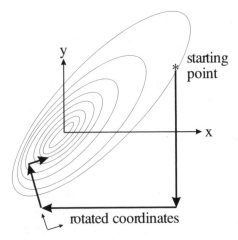

Figure 1.6 Possible path that the Rosenbrock method might take on a quadratic cost surface.

is orthogonal to the previous gradient. If the valley is narrow (ratio of maximum to minimum eigenvalue large), then this algorithm bounces from side to side for many iterations before reaching the bottom. Figure 1.7 shows a possible path of the steepest descent algorithm. Note that the path is orthogonal to the contours and one path is orthogonal to the previous and next path.

Powell developed a method that finds a set of line minimization directions that are linearly independent, mutually conjugate directions (Powell, 1964). The direction assuring the current direction does not "spoil" the

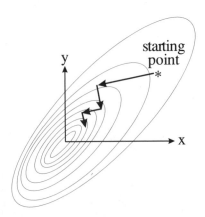

Figure 1.7 Possible path that the steepest descent algorithm might take on a quadratic cost surface.

minimization of the prior direction is the conjugate direction. The conjugate directions are chosen so that the change in the gradient of the cost function remains perpendicular to the previous direction. If the cost function is quadratic, then the algorithm converges in N_{par} iterations (see Figure 1.8). If the cost function is not quadratic, then repeating the N_{par} iterations several times usually brings the algorithm closer to the minimum.

An even better set of directions can be chosen if the matrix of second partial derivatives, the Hessian matrix, is known. The algorithm of Broyden–Fletcher–Goldfarb–Shanno (BFGS), discovered by its four namesakes independently around 1970 (Broyden, 1965; Fletcher, 1963; Goldfarb and Lapidus, 1968; Shanno, 1970), finds a way to approximate this matrix and employs it in determining the appropriate directions of movement. This algorithm is "quasi-Newton" in that it is equivalent to Newton's method for prescribing the next best point to use for the iteration, yet it doesn't use an exact Hessian matrix. The Broyden–Fletcher–Goldfarb–Shanno (BFGS) algorithm is quite robust and quickly converges, but it requires an extra step to approximate the Hessian. Table 1.1 displays the results of minimizing equation (1.1) using the BFGS algorithm. For this "nasty" cost surface, the BFGS algorithm performed slightly worse, but very similar to, the Nelder–Mead algorithm. This algorithm also failed to find the global minimum in 12 runs.

Quadratic programming assumes the cost function is quadratic (parameters are squared) and the constraints are linear. This technique is based upon Lagrange multipliers and requires derivatives or approximations to derivatives. One powerful method known as recursive quadratic programming solves the quadratic programming problem at each iteration to find

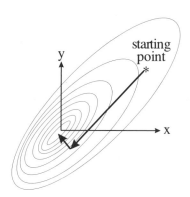

Figure 1.8 Possible path that a conjugate directions algorithm might take on a quadratic cost surface.

the direction of the next step (Luenberger, 1984). The approach of these methods is similar to using very refined surveying tools, which unfortunately still does not help guarantee that the hiker will find the lowest point in the park.

1.3   NATURAL OPTIMIZATION METHODS

The methods already discussed take the same basic approach of heading downhill from an arbitrary starting point. They differ in deciding in which direction to move and how far to move. Successive improvements increased the speed of the downhill algorithms, but didn't add to the algorithm's ability to find a global minimum instead of a local minimum.

All hope is not lost! Some outstanding algorithms have surfaced in the last 30 years. Two relatively new ones are the genetic algorithm and simulated annealing. The genetic algorithm models natural selection and evolution, while simulated annealing models the annealing process. Both methods generate new points in the search space by applying operators to current points and statistically moving toward more optimal places in the search space. Both techniques have met with tremendous success in a number of areas. They rely upon an intelligent search of a large but finite solution space using statistical methods. Both algorithms do not require taking cost function derivatives and can thus deal with discrete parameters and noncontinuous cost functions. They represent processes in nature that are remarkably successful at optimizing natural phenomena.

1.3.1   Simulated Annealing

In the early 1980s the method of simulated annealing was introduced by Kirkpatrick and coworkers (1983), based on ideas formulated in the early 1950s (Metropolis et al., 1953). This method simulates the annealing process in which a substance is heated to a temperature above its melting temperature, then cooled gradually to produce the crystalline lattice, which minimizes its energy probability distribution. This crystalline lattice, composed of millions of atoms perfectly alined, is a beautiful example of nature finding an optimal structure. However, if the cooling proceeds too quickly, that is, is "quenched," the crystal never forms and the substance becomes an amorphous mass with a higher than optimum energy state. The key to crystal formation is carefully controlling the rate of change of temperature. The algorithmic analogue to this process involves initializing a first guess state, then "heating" it by modifying the parameter values. The cost function represents the energy level of the substance. What makes the simulated annealing algorithm unique is the addition of a control parameter, analogous to the temperature, which controls the speed of descent of the algorithm into the optimum cost function value. This control parameter sets the step size, so that at the beginning of the process, the algorithm is forced to make large changes in parameter values. At times, the changes move the algorithm away from the optimum, but it forces the algorithm to explore new regions of parameter space. After a certain number of iterations, the control parameter is lowered, and smaller steps are allowed in the parameters. The control parameter is lowered slowly, so that the algorithm has a chance to find the correct valley before trying to get to the lowest point in the valley. This algorithm has essentially "solved" the traveling salesman problem (Kirkpatrick et al., 1983) and has been applied successfully to a wide variety of problems.

1.3.2 The Genetic Algorithm

A second type of natural method is the genetic algorithm. It is a subset of evolutionary algorithms that model biological processes to optimize highly complex cost functions. A genetic algorithm allows a population composed of many individuals to evolve under specified selection rules to a state that maximizes the "fitness" (i.e., minimizes the cost function). The method was developed by John Holland (1975) over the course of the 1960s and 1970s and finally popularized by one of his students, David Goldberg, who was able to solve a difficult problem involving the control of gas-pipeline transmission for his dissertation (Goldberg, 1989).

Some of the advantages of a genetic algorithm include that it

- Optimizes with continuous or discrete parameters,
- Doesn't require derivative information,
- Simultaneously searches from a wide sampling of the cost surface,
- Deals with a large number of parameters,
- Is well suited for parallel computers,
- Optimizes parameters with extremely complex cost surfaces; they can jump out of a local minimum,
- Provides a list of optimum parameters, not just a single solution,
- May encode the parameters so that the optimization is done with the encoded parameters, and
- Works with numerically generated data, experimental data, or analytical functions.

These advantages are intriguing and produce stunning results when traditional optimization approaches fail miserably.

Of course, the genetic algorithm is not the best way to solve every problem. For instance, the traditional methods have been well-tuned to quickly find the solution of a well-behaved convex analytical function of only a few variables. For such cases, the calculus-based methods outperform the genetic algorithm, quickly finding the minimum while the genetic algorithm is still analyzing the costs of the initial population. For these problems the optimizer should use the experience of the past and employ these quick methods. However, many realistic problems do not fall into this category. In addition, for problems that are not overly difficult, other methods may find the solution faster than the genetic algorithm, at least if one is using a serial computer. The large population of solutions that gives the genetic algorithm its power is also its bane when it comes to speed on a serial computer—the cost function of each of those solutions must be evaluated. However, if a parallel computer is available, each processor can evaluate a separate function at the same time. Thus, the genetic algorithm is optimally suited for such parallel computations.

The remainder of the book more specifically describes the genetic algorithm. Chapter 2 introduces the binary form while using the algorithm to find the highest point in Rocky Mountain National Park. Chapter 3 describes another version of the algorithm that employs continuous parameters. We demonstrate this method with a genetic algorithm solution to equation (1.1) subject to constraints (1.2). The remainder of the book presents refinements to the algorithm by solving more problems, winding its way from easier, less technical problems into more difficult problems that could not be solved by other methods. Our goal is to give specific ways to deal with certain types of problems that may be typical of the ones faced by scientists and engineers on a day-to-day basis. However, before presenting this algorithm in detail in the following chapters, we provide some background to the biological processes of genetics and natural selection in the next section. This is to provide a basis for understanding the roots of the terminology that has invaded the genetic algorithm literature.

1.4 BIOLOGICAL OPTIMIZATION: NATURAL SELECTION

This section introduces the current scientific understanding of the natural selection process with the purpose of gaining an insight into the construction, application, and terminology of genetic algorithms. Natural selection is discussed in many texts and treatises. Much of the information summarized here is from Curtis (1975) and Grant (1985).

Upon observing the natural world, we can make several generalizations that lead to our view of its origins and workings. First, there is a tremendous diversity of organisms. Second, the degree of complexity in the organisms is striking. Third, many of the features of these organisms have an apparent usefulness. Why is this so? How did they come into being?

Imagine the organisms of today's world as being the results of many iterations in a grand optimization algorithm. The cost function measures survivability, which we wish to maximize. Thus, the characteristics of the organisms of the natural world fit into this topological landscape (Grant, 1985). The level of adaptation, the fitness, denotes the elevation of the landscape. The highest points correspond to the most-fit conditions. The environment itself, as well as how the different species interact provide the constraints. The process of evolution is the grand algorithm that selects which characteristics produce a species of organism fit for survival. The peaks of the landscape are populated by living organisms. Some peaks are broad and hold a wide range of characteristics encompassing many organisms, while other peaks are very narrow and allow only very specific characteristics. This analogy can be extended to include saddles between peaks as separating different species. If we take a very parochial view and assume that intelligence and ability to alter the environment are the most important aspects of survivability, we can imagine the global maximum peak at this instance in biological time to contain humankind.

To begin to understand the way that this natural landscape was populated involves studying the two components of natural selection: genetics and evolution. Modern biologists subscribe to what is known as the synthetic theory of natural selection—a synthesis of genetics with evolution. There are two main divisions of scale in this synthetic evolutionary theory: macroevolution, which involves the process of division of the organisms into major groups, and microevolution, which deals with the process within specific populations. We will deal with microevolution in the following discussion and consider macroevolution to be beyond our scope.

First, we need a bit of background on heredity at the cellular level. A *gene* is the basic unit of heredity. An organism's genes are carried on one of a pair of *chromosomes* in the form of deoxyribonucleic acid (DNA). The DNA is in the form of a double helix and carries a symbolic system of base-pair sequences that determine the sequence of enzymes and other proteins in an organism. This sequence does not vary and is known as the *genetic code* of the organism. Each cell of the organism contains the same number of chromosomes. For instance, the number of chromosomes per body cell is 6 for mosquitos, 26 for frogs, 46 for humans, and 94 for goldfish. Genes often occur with two functional forms, each representing a different characteristic. Each of these forms is known as an *allele*. For instance, a human may carry one allele for brown eyes and another for blue eyes. The combination of alleles on the chromosomes determines the traits of the individual. Often one allele is dominant and the other recessive, so that the dominant allele is what is manifested in the organism, although the recessive one may still be passed on to its offspring. If the allele for brown eyes is dominant, the organism will have brown eyes. However, it can still pass the blue allele to its offspring. If the second allele from the other parent is also for blue eyes, the child will be blue-eyed.

The study of genetics began with the experiments of Gregor Mendel. Born in 1822, Mendel attended the University of Vienna, where he studied both biology and mathematics. After failing his exams, he became a monk. It was in the monastery garden where he performed his famous pea plant experiments. Mendel revolutionized experimentation by applying mathematics and statistics to analyzing and predicting his results. By his hypothesizing and careful planning of experiments, he was able to understand the basic concepts of genetic inheritance for the first time, publishing his results in 1865. As with many brilliant discoveries, his findings were not appreciated in his own time.

Mendel's pea plant experiments were instrumental in delineating how traits are passed from one generation to another. One reason that Mendel's experiments were so successful is that pea plants are normally self-pollinating and seldom cross-pollinate without intervention. The self-pollination is easily prevented. Another reason that Mendel's experiments worked was the fact that he spent several years prior to the actual experimentation documenting the inheritable traits and which ones were easily separable and bred pure. This allowed him to crossbreed his plants and observe the characteristics of the offspring and of the next generation. By carefully observing the distribution of traits, he was able to hypothesize his first law—the principle of segregation; that is, that there must be factors that are inherited in pairs, one from each parent. These factors are indeed the genes and their different realizations are the alleles. When both alleles of a gene pair are the same, they are *homozygous*. When they are different, they are *heterozygous*. The brown-blue allele for eye color of a parent was heterozygous while the blue-blue combination of the offspring is homozygous. The trait actually observed is the *phenotype*, but the actual combination of alleles is the *genotype*. Although the parent organism had a brown-blue eye color phenotype, its genotype is for brown eyes (the dominant form). The genotype must be inferred from the phenotype percentages of the succeeding generation as well as the parent itself. Since the offspring had blue eyes, we can insure that each parent had a blue allele to pass along, even though the phenotype of each parent was brown eyes. Therefore, since the offspring was homozygous, carrying two alleles for blue eyes, both parents must be heterozygous, having one brown and one blue allele. Mendel's second law is the principle of independent assortment. This principle states that the inheritance of the allele for one trait is independent of that for another. The eye color is irrelevant when determining the size of the individual.

To understand how genes combine into phenotypes, it is helpful to understand some basics of cell division. Reproduction in very simple, single-celled organisms occurs by cell division, known as *mitosis*. During the phases of mitosis, the chromosome material is exactly copied and passed onto the offspring. In such simple organisms, the daughter cells are identical to the parent. There is little opportunity for evolution of such organisms. Unless a mutation occurs, the species propagates unchanged. Higher organisms have developed a more efficient method of passing on traits to their offspring—sexual reproduction. The process of cell division that occurs then is called *meiosis*. The *gamete*, or reproductive cell, has half the number of chromosomes as the other body cells. Thus the gametes cells are called *haploid*, while the body cells are *diploid*. Only these diploid body cells contain the full genetic code. The diploid number of chromosomes is reduced by half to form the haploid number for the gametes. In preparation for meiosis, the gamete cells are duplicated. Then the gamete cells from the mother join with those from the father (this process is not discussed here). They arrange themselves in *homologous* pairs; that is, each chromosome matches with one of the same length and shape. As they match up, they join at the *kinetochore*, a random point on this matched chromosome pair (or actually tetrad in most cases). As meiosis progresses, the kinetochores divide so that a left portion of the mother chromosome is conjoined with the right portion of the father and visa versa for the other portions. This process is known as *crossing over*. The resulting cell has the full diploid number of chromosomes. Through this crossing over, the genetic material of the mother and father have been combined in a manner to produce a unique individual offspring. This process allows changes to occur in the species.

Now we turn to discussing the second component of natural selection— evolution—and one of its first proponents, Charles Darwin. Darwin refined his ideas during his voyage as naturalist on the *Beagle*, especially during his visits to the Galapagos Islands. Darwin's theory of evolution was based on four primary premises. First, like begets like; equivalently, an offspring has many of the characteristics of its parents. This premise implies that the population is stable. Secondly, there are variations in characteristics between individuals which can be passed from one generation to the next. The third premise is that only a small percentage of the offspring produced survive to adulthood. Finally, which of the offspring survive depends on their inherited characteristics. These premises combine to produce the theory of natural selection. In modern evolutionary theory, an understanding of genetics adds impetus to the explanation of the stages of natural selection.

A group of interbreeding individuals is called a *population*. Under static conditions, the characteristics of the population are defined by the *Hardy-Weinberg Law*. This principle states that the frequency of occurrence of the alleles will stay the same within an inbreeding population if there are no perturbations. Thus, although the individuals show great variety, the statistics of the population remain the same. However, we know that few populations are static for very long. When the population is no longer static, the proportion of allele frequencies is no longer constant between generations and evolution occurs. This dynamic process requires an external forcing. The forcing may be grouped into four specific types. (1) *Mutations* may occur; that is, a random change occurs in the characteristics of a gene. This change may be passed along to the offspring. Mutations may be spontaneous or due to external factors such as exposure to environmental factors. (2) *Gene flow* may result from introduction of new organisms into the breeding population. (3) *Genetic drift* may occur solely due to chance. In small populations, certain alleles may sometimes be eliminated in the random combinations. (4) *Natural selection* operates to choose the *most fit* individuals for further reproduction. In this process, certain alleles may produce an individual that is more prepared to deal with its environment. For instance, fleeter animals may be better at catching prey or running from predators, thus being more likely to survive to breed. Therefore, certain characteristics are *selected* into the breeding pool.

Thus we see that these ideas return to natural selection. The important components have been how the genes combine and cross over to produce new individuals with combinations of traits and how the dynamics of a large population interact to select for certain traits. These factors may move this offspring either up toward a peak or down into the valley. If it goes too far into the valley, it may not survive to mate—better adapted ones will. After a long period of time, the pool of organisms becomes well adapted to its environment. However, the environment is dynamic. The predators and prey, as well as factors such as the weather and geological upheaval are also constantly changing. These changes act to revise the optimization equation. That is what makes life (and genetic algorithms) interesting.

BIBLIOGRAPHY

Anderson, D. Z., 1992, "Linear programming," in *McGraw-Hill Encyclopedia of Science and Technology* **10**, New York: McGraw-Hill, pp. 86–88.

Borowski, E. J., and J. M. Borwein, *Mathematics Dictionary*, 1991, New York: HarperCollins.

Box, M. J., 1965, "A comparison of several current optimization methods and the use of transformations in constrained problems," *Comput. J.* **8**, pp. 67–77.

Boyer, C. B., and U. C. Merzbach, 1991, *A History of Mathematics*, New York: John Wiley.

Broyden, G. C., 1965 Oct., "A class of methods for solving nonlinear simultaneous equations," *Math. Comput.*, pp. 577–593.

Curtis, H., 1975, *Biology*, 2nd Ed., New York: Worth Publishers.

Cuthbert, T. R., Jr., 1987, *Optimization Using Personal Computers*, New York: John Wiley.

Fletcher, R., 1963, "Generalized inverses for nonlinear equations and optimization," *Numerical Methods for Non-linear Algebraic Equations* (R. Rabinowitz, Ed.), London: Gordon & Breach.

Goldberg, D. E., 1989, *Genetic Algorithms in Search, Optimization, and Machine Learning*, New York: Addison-Wesley.

Goldfarb, D., and B. Lapidus, 1968 Feb., "Conjugate gradient method for nonlinear programming problems with linear constraints," *I&EC Fundam.*, pp. 142–151.

Grant, V., 1985, *The Evolutionary Process*, New York: Columbia University Press.

Holland, J. H., 1975, *Adaptation in Natural and Artificial Systems*, Ann Arbor: The University of Michigan Press.

Kirkpatrick, S., C. D. Gelatt, Jr., and M. P. Vecchi, 1983 May 13, "Optimization by simulated annealing," *Science* **220**, pp. 671–680.

Luenberger, D. G., 1984, *Linear and Nonlinear Programming*, Reading, MA: Addison-Wesley, 1984.

Metropolis, N., et al., 1953, "Equation of state calculations by fast computing machines," *J. Chem. Phys.* **21**, pp. 1087–1092.

Nelder, J. A., and R. Mead, 1965, "A simplex method for function minimization," *Comput. J.* **7**, pp. 308–313.

Pierre, D. A., 1992, "Optimization," in *McGraw-Hill Encyclopedia of Science and Technology* **12**, New York: McGraw-Hill, pp. 476–482.

Powell, M. J. D., 1964, "An efficient way for finding the minimum of a function of several variables without calculating derivatives," *Comput. J.* **7**, pp. 155–162.

Press, W. H., et al., 1992, *Numerical Recipes*, New York: Cambridge University Press.

Rosenbrock, H. H., 1960, "An automatic method for finding the greatest or least value of a function," *Comput. J.* **3**, pp. 175–184.

Schwefel, H., 1995, *Evolution and Optimum Seeking*, New York: John Wiley.

Shanno, D. F., 1970 March, "An accelerated gradient projection method for linearly constrained nonlinear estimation," *SIAM J. Appl. Math.*, pp. 322–334.

Thompson, G. L., 1992, "Game theory," in *McGraw-Hill Encyclopedia of Science and Technology* **7**, New York: McGraw-Hill, pp. 555–560.

Williams, H. P., 1993, *Model Solving in Mathematical Programming*, New York: John Wiley.

CHAPTER 2

THE BINARY GENETIC ALGORITHM

2.1 GENETIC ALGORITHMS: NATURAL SELECTION ON A COMPUTER

The previous chapter whet your appetite for something better than the traditional optimization methods. This and the next chapter give step-by-step procedures for implementing two flavors of a genetic algorithm. Both algorithms follow the same menu of modeling genetic recombination and natural selection. One represents parameters as an encoded binary string and works with the binary strings to minimize the cost, while the other works with the continuous parameters themselves to minimize the cost. Since genetic algorithms originated with a binary representation of the parameters, it is presented first.

Figure 2.1 shows the analogy between biological evolution and a binary genetic algorithm. Both start with an initial population of random members. On the left side, each row of binary numbers represents characteristics of one of the dogs in the population. If we are trying to breed the dog with the loudest bark, then only a few of the best (in this case, four best) barking dogs are kept for breeding. Traits associated with loud barking are encoded in the binary sequence associated with these dogs. From this breeding population, two are randomly selected to create two new puppies. The puppies have a high probability of being loud barkers, because both

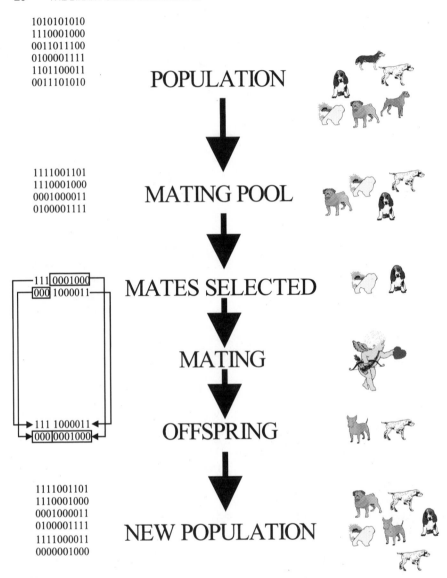

Figure 2.1 Analogy between a numerical genetic algorithm and biological genetics.

their parents have genes that make them loud barkers. The new binary sequences of the puppies contain portions of the binary sequences of both parents. These new puppies replace two discarded dogs that didn't bark loud enough. Enough puppies are generated to bring the population back to its original size. Iterating on this process leads to a dog with a very loud bark.

2.2 COMPONENTS OF A BINARY GENETIC ALGORITHM

The genetic algorithm begins, like any other optimization algorithm, by defining the optimization parameters, the cost function, and the cost. It ends like other optimization algorithms too, by testing for convergence. In between, however, this algorithm is very different from other optimization algorithms. A path through the components of the genetic algorithm is shown as a flow chart in Figure 2.2. Each block in this "big picture" overview is discussed in detail in this chapter.

In the previous chapter, the cost function was a surface with peaks and valleys when displayed in parameter space, much like a topographic map. To find a valley, an optimization algorithm searches for the minimum cost.

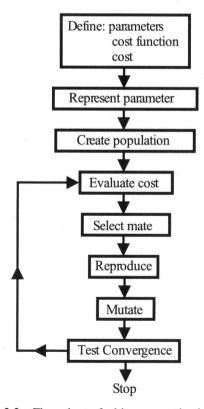

Figure 2.2   Flow chart of a binary genetic algorithm.

To find a peak, an optimization algorithm searches for the maximum cost. This analogy leads to the problem of finding the highest point in Rocky Mountain National Park. A three dimensional plot of a portion of the park (our search space) is shown in Figure 2.3 and a crude topographical map (128 × 128 points) with some of the highlights is shown in Figure 2.4. Locating the top of Long's Peak (14,255 ft above sea level) is the goal. Three other interesting features in the area include Storm Peak (13,326 ft), Mount Lady Washington (13,281 ft), and Chasm Lake (11,800 ft). Since there are many peaks in the area of interest, conventional optimization techniques have difficulty finding Long's Peak unless the starting point is in the immediate vicinity of the peak. In fact, all of the methods requiring a gradient of the cost function won't work with discrete data. The genetic algorithm has no problem!

2.2.1 Selecting the Parameters and the Cost Function

A cost function generates an output from a set of input parameters (a chromosome). The cost function may be a mathematical function, an experiment, or a game. The object is to modify the output in some desirable fashion by finding the appropriate values for the input parameters. We do

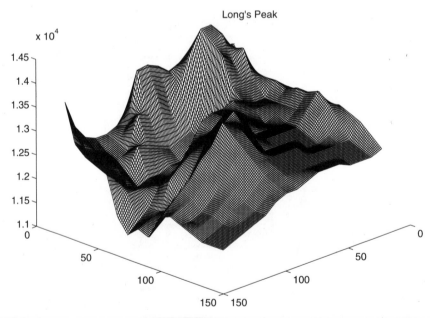

Figure 2.3 Three-dimensional view of the cost surface with a view of Long's Peak.

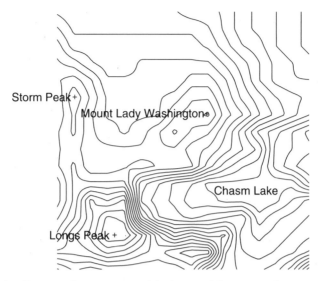

Figure 2.4   Contour plot or topographical map of the cost surface around Long's Peak.

this without thinking when filling a bathtub with water. The cost is the difference between the desired and actual temperatures of the water. The input parameters are how much the hot and cold spigots are turned. In this case, the cost function is the experimental results of feeling the resulting temperature. So we see that determining an appropriate cost function and deciding which parameters to use are intimately related.

The genetic algorithm begins by defining a chromosome or an array of parameter values to be optimized. If the chromosome has N_{par} parameters (an N_{par}-dimensional optimization problem) given by $p_1, p_2, \ldots, p_{N_{par}}$, then the chromosome is written as an N_{par} element array.

$$chromosome = [p_1, p_2, p_3, \ldots, p_{N_{par}}] \tag{2.1}$$

For instance, searching for the maximum elevation on a topographical map requires a cost function with input parameters of longitude (x) and latitude (y)

$$chromosome = [x, y] \tag{2.2}$$

where $N_{par} = 2$. Each chromosome has a cost found by evaluating the cost function, f, at $p_1, p_2, \ldots, p_{N_{par}}$:

$$cost = f(chromosome) = f(p_1, p_2, \ldots, p_{N_{par}}) \tag{2.3}$$

Since we are trying to find the peak in Rocky Mountain National Park, the cost function is written as the negative of the elevation in order to put it into the form of a minimization algorithm:

$$f(x, y) = -elevation \tag{2.4}$$

Often, the cost function is quite complicated, as in maximizing the gas mileage of a car. The user must decide which parameters of the problem are most important. Too many parameters bog down the genetic algorithm. Important parameters for optimizing the gas mileage might include size of the car, size of the engine, and weight of the materials. Other parameters, such as paint color and type of headlights, have little or no impact on the car gas mileage and should not be included. Sometimes the correct number and choice of parameters comes from experience or trial optimization runs. Other times we have an analytical cost function, with the parameter being the variables of the function. A cost function defined by $f(w, x, y, z) = 2x + 3y + z/100000 + \sqrt{w}/9876$ with all parameters lying between 1 and 10 can be simplified to help the optimization algorithm. Since the w and z terms are extremely small in the region of interest, they can be discarded for most purposes. Thus, the four-dimensional cost function is adequately modeled with two parameters in the region of interest.

Most optimization problems require constraints or parameter bounds. Allowing the weight of the car to go to zero or letting the car width be 10 m are impractical parameter values. Unconstrained parameters can take any value. Constrained parameters come in three brands. First, hard limits in the form of $>$, $<$, $\geq$, and $\leq$ can be imposed on the parameters. When a parameter exceeds a bound, then it is set equal to that bound. If x has limits of $0 \leq x \leq 10$, and the algorithm assigns $x = 11$, then x will be reassigned to the value of 10. Second, variables can be transformed into new variables that inherently include the constraints. If x has limits of $0 \leq x \leq 10$, then $x = 5 \sin y + 5$ is a transformation between the constrained variable x and the unconstrained variable y. Varying y for any value is the same as varying x within its bounds. This type of transformation changes a constrained optimization problem into an unconstrained optimization problem in a smooth manner. Finally, there may be a finite set of parameter values from which to choose, and all values lie within the region of interest. Such problems come in the form of selecting parts from a limited supply.

Dependent parameters present special problems for optimization algorithms, because varying one parameter also changes the value of the other parameter. For example, size and weight of the car are dependent. Increasing the size of the car will most likely increase the weight as well (unless some other factor, such as type of material, is also changed). Independent parameters, like Fourier series coefficients, do not interact with each other. If 10 coefficients are not enough to represent a function, then more can be added without having to recalculate the original ten.

In the genetic algorithm literature, parameter interaction is called *epistasis* (a biological term for gene interaction). When there is little to no epistasis, minimum seeking algorithms perform best. Genetic algorithms shine when the epistasis is medium to high, and pure random search algorithms are champions when epistasis is very high (Figure 2.5).

Defining the cost is a challenge. Should the cost only be related to one observable output, or should it be a combination of outputs? Let's look at the situation of the car design. Is letting $cost = $ mi/gal a good enough cost function for the design of a car? Assume the money needed to build the car ($) and the gas mileage ($M$) are measures of the car design with parameters being weight (w) and volume (v) (We wouldn't buy this car!). An example cost function is

$$cost(w, v) = \$ + M \qquad (2.5)$$

If the cost is about $10,000 and the mileage 50 mi/gal, then the cost function is skewed toward the monetary cost of the vehicle. If both monetary cost and mileage are equally important, then the cost can be normalized as

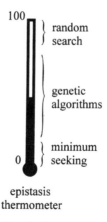

epistasis
thermometer

Figure 2.5 This graph of an epistasis thermometer shows that minimum seeking algorithms work best for low epistasis, while random algorithms work best for very high epistasis. Genetic algorithms work best in a wide range of medium to high epistasis.

follows:

$$cost = \frac{\$ - \$_{lo}}{\$_{hi} - \$_{lo}} + \frac{M - M_{lo}}{M_{hi} - M_{lo}} \tag{2.6}$$

where the *hi* and *lo* subscripts indicate the maximum and minimum values, respectively. This cost lies between 0 and 2, with the *lo* and *hi* subscripts representing the low and high values of the parameters. If one portion of the cost is more important, then it can be appropriately weighted ($0 < wt < 1$):

$$cost = wt\frac{\$ - \$_{lo}}{\$_{hi} - \$_{lo}} + (1 - wt)\frac{M - M_{lo}}{M_{hi} - M_{lo}} \tag{2.7}$$

The genetic algorithm is sensitive to the parameter range and cost representation. Equation (2.7) allows the user excellent control over the desired outcome. Picking a good *wt* is not always easy though.

2.2.2 Parameter Representation

The binary genetic algorithm works with a finite (but usually extremely large) parameter space. This characteristic makes the genetic algorithm ideal for optimizing a cost that is due to parameters that can only assume a finite number of values. Common examples include choosing values from a list or parts from stock on hand.

If a parameter is continuous, then it must be quantized. Quantizing a parameter or signal is an established art. First, the range is divided into equal quantization levels (Figure 2.6). Any value falling within one of the levels is set equal to the mid, high, or low value of that level. In general, setting the value to the mid value of the quantization level is best, because the largest error possible is half a level. Rounding the value to the low or high value of the level allows a maximum error equal to the quantization level. The mathematical formulas for the binary encoding and decoding of the *n*th parameter, p_n, are given by
    encoding:

$$p_{norm} = \frac{p_n - p_{lo}}{p_{hi} - p_{lo}} \tag{2.8}$$

$$gene[m] = \textbf{round}\left\{ p_{norm} - 2^{-m} - \sum_{p=1}^{m-1} gene[m]2^{-p} \right\} \tag{2.9}$$

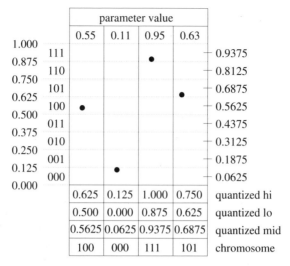

Figure 2.6   Four continuous parameter values are graphed with the quantization levels shown. The corresponding gene or chromosome indicates the quantization level where the parameter value falls. Each chromosome corresponds to a low, mid, or high value in the quantization level. Normally, the parameter is assigned the mid value of the quantization level.

decoding:

$$p_{quant} = \sum_{m=1}^{N_{gene}} gene[m]2^{-m} + 2^{-(M+1)} \tag{2.10}$$

$$q_n = p_{quant}(p_{hi} - p_{lo}) + p_{lo} \tag{2.11}$$

where

$$
\begin{aligned}
p_{norm} &= \text{normalized parameter, } 0 \le p_{norm} \le 1 \\
p_{lo} &= \text{smallest parameter value} \\
p_{hi} &= \text{highest parameter value} \\
gene[m] &= \text{binary version of } p_n \\
\mathbf{round}\{\cdot\} &= \text{round to nearest integer} \\
N_{gene} &= \text{number of bits in the gene} \\
p_{quant} &= \text{quantized version of } p_{norm} \\
q_n &= \text{quantized version of } p_n
\end{aligned}
$$

The genetic algorithm works with the binary encodings, but the cost function often requires continuous parameters. Whenever the cost function is evaluated, the chromosome must first be decoded using equation (2.10). An example of a binary encoded chromosome that has N_{par} parameters, each encoded with $N_{gene} = 10$ bits, is

$$chromosome = \left[\underbrace{1111001001}_{gene_1} \underbrace{0011011111}_{gene_2} \cdots \underbrace{0000101001}_{gene_{N_{par}}} \right]$$

Substituting each gene in this chromosome into equation (2.10) yields an array of the quantized version of the parameters. This chromosome has a total of $N_{bits} = N_{gene} \times N_{par} = 10 \times N_{par}$ bits.

As previously mentioned, the topographical map of Rocky Mountain National Park has 128×128 elevation points. If x and y are encoded in two genes, each with $N_{gene} = 7$ bits, then there are 2^7 possible values for x and y. These values range from $40° 15' \leq y \leq 40° 16'$ and $105° 37'30'' \geq x \geq 105° 36'$. Thus, a random chromosome may have the following $N_{bits} = 14$-bit binary representation:

$$chromosome = \left[\underbrace{1100011}_{x} \underbrace{0011001}_{y} \right]$$

This chromosome translates into matrix coordinates of $[99, 25]$ or longitude, latitude coordinates of $[105° 36'50'', 40° 15'29.7'']$.

2.2.3   Initial Population

The genetic algorithm starts with a large commune of chromosomes known as the initial population. This initial population has N_{ipop} chromosomes and is an $N_{ipop} \times N_{bits}$ matrix filled with random ones and zeros generated from

$$IPOP = \mathbf{round}\{\mathbf{random}(N_{ipop}, N_{bits})\}$$

where the function $\mathbf{random}(N_{ipop}, N_{bits})$ generates an $N_{ipop} \times N_{bits}$ matrix of uniform random numbers between zero and one. This type of function is available on all standard mathematical software. The function $\mathbf{round}\{\}$ rounds the numbers to the closest integer. Each row in the matrix is a chromosome. The chromosomes correspond to discrete values of longitude and latitude. Next, the parameters are passed to the cost function for evaluation. A large initial population provides the genetic algorithm with a nice sampling of the search space. Usually, not all the initial population matrix chromosomes make the cut for the iterative portion of the genetic algorithm. Table 2.1 shows the initial population and their costs for the

TABLE 2.1   Example Initial Population of 24 Random Chromosomes and Their Corresponding Cost

Chromosome	Cost
00000000000000	-13000
11111011010010	-11800
00010110000010	-13255
11000011001010	-12347
01111111101001	-12560
01000111010001	-12700
01010110000100	-13338
11101111001110	-11890
01111100111100	-12953
00100001011110	-12891
10001110111010	-12759
10111000111100	-12320
11011011101000	-11797
00100110011101	-13778
00010100011011	-13360
01110010101011	-12220
11000011001100	-12452
10011101110000	-12335
10100000000011	-12857
00001101010110	-13166
00010000110101	-13164
01101100110010	-12927
01101111000010	-13079
10001001011111	-12756

generation 1

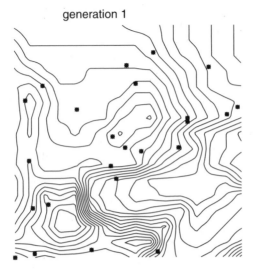

Figure 2.7   A contour map of the cost surface with the 24 initial population members indicated by asterisks in circles.

N_{ipop} = 24 random chromosomes. The locations of the chromosomes are shown on the topographical map in Figure 2.7.

2.2.4   Natural Selection

The initial population is too large to undergo the journey through the iterative steps of the genetic algorithm. Thus, a large portion of the high cost chromosomes are discarded through natural selection or survival of the fittest. First, the N_{ipop} costs and associated chromosomes are ranked from lowest cost to highest cost. Then, only the best $N_{pop} \leq N_{ipop}$ members of the population are kept for each iteration of the genetic algorithm, while the others are discarded. The authors have found that letting $N_{ipop} > N_{pop}$ gives the genetic algorithm a nice start by providing an excellent initial sampling of the cost surface. Natural selection occurs each generation or iteration of the algorithm. Of the N_{pop} chromosomes in a generation, only the top N_{good} survive for mating, and the bottom N_{bad} are discarded to make room for the new offspring.

Deciding how many chromosomes to keep is somewhat arbitrary. Letting only a few chromosomes survive to the next generation limits the available genes in the offspring. Keeping too many chromosomes allows bad performers a chance to contribute their traits to the next generation. We often keep 50% in the natural selection process.

In our example, $N_{ipop} = 24$. Only $N_{pop} = 12$ of these chromosomes are kept for the population each iteration of the genetic algorithm. Every iteration, $N_{good} = 6$ chromosomes are placed in a mating pool for reproducing and $N_{bad} = 6$ are discarded. The natural selection results are shown in Table 2.2. Note that the chromosomes of Table 2.2 have first been sorted by cost. Then, the six with the lowest cost survive to the next generation and become potential parents.

Another approach to natural selection is called *thresholding*. In this approach, all chromosomes that have a cost lower than some threshold survive. The threshold must allow some chromosomes to continue in order to have parents to produce offspring. Otherwise a whole new initial population must be generated to find some chromosomes that pass the test. At

TABLE 2.2   Ranking and Categorization of Initial Population of 24 Random Chromosomes

Chromosome	Cost
N_{ipop} { N_{good} { 00100110011101	-13778 N_{pop}
00010100011011	-13360
01010110000100	-13338
00010110000010	-13255
00001101010110	-13166
00010000110101	-13164
N_{bad} { 01101111000010	-13079
00000000000000	-13000
01111100111100	-12953
01101100110010	-12927
00100001011110	-12891
10100000000011	-12857
10001110111010	-12759
10001001011111	-12756
01000111010001	-12700
01111111101001	-12560
11000011001100	-12452
11000011001010	-12347
10011101110000	-12335
10111000111100	-12320
01110010101011	-12220
11101111001110	-11890
11111011010010	-11800
11011011101000	-11797

first, only a few chromosomes may survive. In later generations, however, most of the chromosomes will survive unless the threshold is changed. An attractive feature of this technique is that the population does not have to be sorted.

2.2.5 Pairing

Now it's time to play matchmaker. Two chromosomes are selected from the mating pool of N_{good} chromosomes to produce two new offspring. Pairing takes place in the mating population until N_{bad} offspring are born to replace the discarded chromosomes. Pairing chromosomes in a genetic algorithm can be as interesting and varied as pairing in an animal species. We'll look at a variety of methods, starting with the easiest.

1. Pairing from top to bottom. Start at the top of the list and pair the chromosomes two at a time until the top N_{good} chromosomes are selected for mating. Thus, the algorithm pairs $chromosome_{2i-1}$ with $chromosome_{2i}$ for $i = 1, 2, \ldots$. In our example, we would pair chromosomes 1&2, 3&4, and 5&6. This approach doesn't model nature well but is very simple to program. It's a good one for beginners to try.

2. Random pairing. This approach uses a uniform random number generator to select chromosomes. The chromosomes are ranked in terms of cost from 1 to N_{good} and two random numbers are generated to find the first two mates. A parent is selected by

$$parent = \textbf{roundup}\{N_{good} \times \textbf{random}\} \tag{2.12}$$

where **roundup**{ } rounds its argument to the next highest integer. For example, six uniform random numbers are generated: 0.1535, 0.6781, 0.0872, 0.1936, 0.7021, and 0.3933. Multiplying these random numbers by six and rounding up gives the following random integer values corresponding to the selected chromosomes: 1, 5, 1, 2, 5, and 3. Thus, $chromosome_1$ is paired with $chromosome_5$, $chromosome_1$ is paired with $chromosome_2$, and $chromosome_5$ is paired with $chromosome_3$.

3. Weighted random pairing. This approach assigns probabilities to the chromosomes in the mating pool according to their cost function. A chromosome with the lowest cost function has the greatest probability of mating, while the chromosome with the highest cost function has the lowest probability of mating. A random number determines which chromosome is selected. This type of weighting is often referred to as roulette wheel weighting. There are two techniques: rank weighting and cost weighting.

TABLE 2.3 Mating Probabilities Assigned According to the Normalized Cost

n	Chromosome	P_n	$\sum_{i=1}^{n} P_i$
1	00100110011101	0.2857	0.2857
2	00010100011011	0.2381	0.5238
3	01010110000100	0.1905	0.7143
4	00010110000010	0.1429	0.8572
5	00001101010110	0.0952	0.9524
6	00010000110101	0.0476	1.0000

(a) Rank weighting. An approach that is problem independent finds the probability from the rank of the chromosome:

$$P_n = \frac{N_{good} - n + 1}{\sum_{n=1}^{N_{good}} n} = \frac{6 - n + 1}{1 + 2 + 3 + 4 + 5 + 6} = \frac{7 - n}{21} \qquad (2.13)$$

Table 2.3 shows the results for the N_{good} chromosomes of our example. The cumulative probabilities listed in column 4 are used in selecting the chromosome. A random number between zero and one is generated. Starting at the top of the list, the first chromosome with a cumulative probability that is greater than the random number is selected for the mating pool. Using the same six random numbers as with the random pairing (0.1535, 0.6781, 0.0872, 0.1936, 0.7021, and 0.3933) gives: *chromosome*$_1$ is paired with *chromosome*$_3$, *chromosome*$_1$ is paired with *chromosome*$_1$, and *chromosome*$_3$ is paired with *chromosome*$_2$. If a chromosome is paired with itself, there are a number of alternatives. First, let it go. It just means there are three of these chromosomes in the next generation. Second, randomly pick another chromosome. The randomness in this approach is more indicative of nature. Rank weighting is only slightly more difficult to program than the pairing from top to bottom.

(b) Cost weighting. The probability is calculated from the cost of the chromosome. A normalized cost is calculated for each chromosome by subtracting the lowest cost of the discarded chromosomes ($cost_{N_{good}+1}$) from the cost of all the chromosomes in the mating pool:

$$C_n = cost_n - cost_{N_{good}+1} \qquad (2.14)$$

TABLE 2.4   Ranking of Chromosomes in the Mating Pool According to Their Costs*

n	Chromosome	$C_n = c_n - c_{N_{good}+1}$	P_n	$\sum_{i=1}^{n} P_i$
1	00100110011101	$-13778 + 13079 = -699$	0.4401	0.4401
2	00010100011011	$-13360 + 13079 = -281$	0.1772	0.6174
3	01010110000100	$-13338 + 13079 = -259$	0.1632	0.7805
4	00010110000010	$-13255 + 13079 = -176$	0.1109	0.8915
5	00001101010110	$-13166 + 13079 = -87$	0.0547	0.9461
6	00010000110101	$-13164 + 13079 = -85$	0.0539	1.0000

*Columns 4 and 5 show two possible mating probabilities.

Subtracting $cost_{N_{good}+1}$ insures all the costs are negative. Table 2.4 lists the normalized costs assuming that $cost_{N_{good}+1} = -13079$. P_n is calculated from

$$P_n = \left| \frac{C_n}{\sum_{p=1}^{N_{good}} C_p} \right| \qquad (2.15)$$

This approach tends to weight the top chromosome more when there is a large spread in the cost between the top and bottom chromosome. On the other hand, it tends to weight the chromosomes evenly when all the chromosomes have approximately the same cost. Using the same six random numbers as before and column 5 of Table 2.4, $chromosome_1$ is paired with $chromosome_3$, $chromosome_1$ is paired with $chromosome_1$, and $chromosome_3$ is paired with $chromosome_1$.

4. Tournament selection. Another approach that closely mimics mating competition in nature is to randomly pick a small subset of chromosomes from the mating pool, and the chromosome with the lowest cost in this subset becomes a parent. The tournament repeats for every parent needed. Consider having six tournaments between two randomly selected chromosomes from the mating pool and selecting the best chromosomes from the tournaments as parents. Results are shown in Table 2.5. Thresholding and tournament selection make a nice pair, because the population never needs to be sorted. Using tournament selection $chromosome_1$ is paired with $chromosome_5$, $chromosome_3$ is paired with $chromosome_4$, and $chromosome_1$ is paired with $chromosome_4$.

Each of the parent selection schemes resulted in a different set of parents. As such, the composition of the next generation is different for each selection scheme. We primarily use rank and cost weighting. Sometimes one works better than the other. It is very difficult to give advice on which

TABLE 2.5 Subset Generations of Two Random Integers Between 1 and 6*

Tournament	$6 \times \text{random}(2, 1)$	Parent
1	2, 1	1
2	5, 5	5
3	6, 3	3
4	4, 5	4
5	1, 1	1
6	4, 5	4

*Each corresponds to selected chromosomes; the chromosome with the lowest cost becomes the parent.

weighting scheme works best. In this example we follow the cost weighting parent selection procedure.

2.2.6 Mating

Mating is the creation of one or more offspring from the parents selected in the pairing process. It is the first way a genetic algorithm explores a cost surface. This is called *exploration* because the genetic algorithm makes use of the bit combinations already present in the chromosomes. The genetic makeup of the population is limited by the current members of the population and helps the genetic algorithm converge. The most common form of mating involves two parents that produce two offspring. A crossover point is selected between the first and last bits of the parents' chromosomes. First, *parent*$_1$ passes its binary code to the left of that crossover point to *offspring*$_1$. In a like manner, *parent*$_2$ passes its binary code to the left of the same crossover point to *offspring*$_2$. Next, the binary code to the right of the crossover point of *parent*$_1$ goes to *offspring*$_2$ and *parent*$_2$ passes its code to *offspring*$_1$. Consequently, the offspring contain portions of the binary codes of both parents. The parents have produced a total of N_{bad} offspring, so the chromosome population is back to N_{pop}. Table 2.6 shows the pairing and mating process for the problem at hand. This process is known as simple or single-point crossover. More complicated versions are discussed in Chapter 5.

2.2.7 Mutations

Random mutations alter a small percentage of the bits in the list of chromosomes. Mutations are the second way a genetic algorithm explores a

TABLE 2.6 Pairing and Mating Process of Single-Point Crossover*

Chromosome	Family	Binary String	Cost
1	$parent_1$	00100110011 101	−13778
3	$parent_2$	01010110000 100	−13338
7	$offspring_1$	00100110011 100	−13372
8	$offspring_2$	01010110000101	−13563
1	$parent_3$	001001100 11101	−13778
1	$parent_4$	001001100 11101	−13778
9	$offspring_3$	001001100 11101	−13778
10	$offspring_4$	00100110011101	−13778
3	$parent_5$	0 1010110000100	−13338
1	$parent_6$	0 0100110011101	−13778
11	$offspring_5$	0 0100110011101	−13778
12	$offspring_6$	01010110000100	−13338

*Two parents produce 2 offspring; parents and offspring become members of the next generation.

cost surface. It can introduce traits not in the original population and keeps the genetic algorithm from converging too fast. A single point mutation changes a "1" to a "0" or visa versa. Mutation points are randomly selected from the $N_{pop} \times N_{bits}$ total number of bits in the population matrix. Increasing the number of mutations increases the algorithm's freedom to search outside the current region of parameter space. It also tends to distract the algorithm from converging on a solution. Typically, on the order of 1% to 5% of the bits mutate per iteration. Mutations do not occur on the final iteration. Do we also allow mutations on the best solutions? Generally not. They are designated as *elite* solutions destined to propagate unchanged. Such elitism is very common in genetic algorithms. Why throw away a perfectly good answer? Other optimization techniques would do well to adopt the elitist strategy.

We want to mutate 5% of the population ($\mu = 0.05$) except for the best chromosome. Thus, a random number generator is used to find seven pairs of random integers that correspond to the rows and columns of the mutated bits. The first random pair is (4, 11). Thus, the bit in row 4 and column 11 is mutated from a *0* to a *1*:

$$00010110000010 \Rightarrow 0001011000\mathit{1}010 \qquad (2.16)$$

Mutations occur six more times at the random rows and columns given by (9, 3), (2, 2), (2, 1), (5, 14), (8, 10), and (5, 8). Most mutations raise the cost of a chromosome. The occasional lowering of the cost adds diversity and strengthens the population. Mutation provides for exploration of the cost surface, because it strays away from the convergence path into new territory.

2.2.8 The Next Generation

After the mutations take place, the costs associated with the offspring and mutated chromosomes are calculated. The process described is iterated. For our example, the next generation is shown in Table 2.7. The italicized digits are mutated bits. The next step ranks them and selects the mating pool, as

TABLE 2.7 Formation of Costs Associated with Offspring and Mutated Chromosomes*

Chromosome	Cost
00100110011101	−13778
*1*1010100011011	−11956
01010110000100	−13338
0001011000*1*010	−13553
00001100010*11*1	−13289
00010000110101	−13164
00100110000100	−13372
0101011000*1*101	−13632
000*0*00110011101	−13036
00100110011101	−13778
00100110011101	−13778
01010110000100	−13338

*Generation 2 is formed after the first generation discards weakest members, mates, and mutates. Mutations are indicated by emphasized bits.

TABLE 2.8   Ranking of Generation 2 from Least
to Most Cost

Chromosome	Cost
00100110011101	−13778
00100110011101	−13778
00100110011101	−13778
01010110001101	−13632
00010110001010	−13552
00100110000100	−13372
01010110000100	−13338
01010110000100	−13338
00001100010111	−13289
00010000110101	−13164
00000110011101	−13036
11010100011011	−11956

shown in Table 2.8. The mean of the first generation was −12738. This
second generation has a mean of −13334. Figure 2.8 shows the location of
these 12 chromosomes. Only the top six are kept for the mating pool. After

generation 2

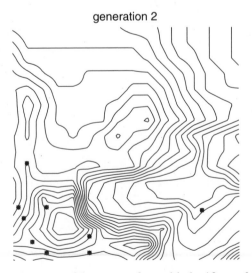

Figure 2.8   A contour map of the cost surface with the 12 members of the second
generation indicated by asterisks in circles.

TABLE 2.9   Ranking of the Third Generation

Chromosome	Cost
00100110011101	−13778
00100110011101	−13778
00100110011101	−13778
00100110011101	−13778
01010110001101	−13632
01010110001101	−13632
00010110001100	−13584
00010110001010	−13553
01010110001010	−13539
00100111010100	−12921
01101111011101	−12602
01100110011101	−12255

mating, mutation, and ranking, the third generation is listed in Table 2.9 (Figure 2.9). The mean is −13403. The fourth generation (Figure 2.10) has a mean of −13676. The means have consistently gotten smaller (larger in magnitude) each generation.

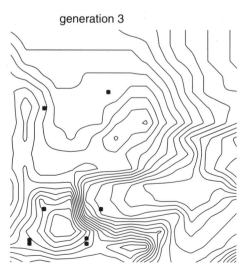

generation 3

Figure 2.9   A contour map of the cost surface with the 12 members of the third generation indicated by asterisks in circles.

generation 4

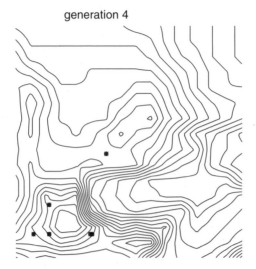

Figure 2.10 A contour map of the cost surface with the 12 members of the fourth generation indicated by asterisks in circles.

2.2.9 Convergence

The number of generations that evolve depends on whether an acceptable solution is reached or a set number of iterations is exceeded. After a while, all the chromosomes and associated costs would become the same if it were not for mutations. At this point, the algorithm should be stopped.

Most genetic algorithms keep track of the population statistics in the form of population mean, standard deviation, and minimum cost. Any of these or any combination of these can serve as a convergence test. For our example, after nine iterations the global minimum is found to be -14199. This minimum was found in less than

$$\underset{\substack{\text{initial population}}}{24} + \underset{\substack{\text{offspring}}}{6} \times \underset{\substack{\text{generations}}}{8} + \underset{\substack{\text{mutated parents}}}{7} \times \underset{\substack{\text{generations}}}{8} = 128 \quad (2.17)$$

cost function evaluations or checking $128/(128 \times 128) \times 100 = 0.78\%$ of the population. The final population is shown in Figure 2.11, where the members are noticeably clumped around Long's Peak. Figure 2.12 shows a plot of the algorithm convergence in terms of the minimum and mean cost of each generation. Long's Peak is actually 14,255 ft above sea level, but the quantization error produced a maximum of 14,199.

generation 8

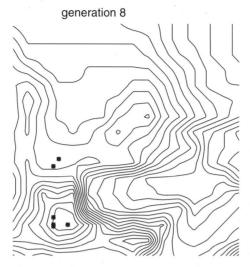

Figure 2.11 A contour map of the cost surface with the final 12 members (ninth generation) indicated by asterisks in circles.

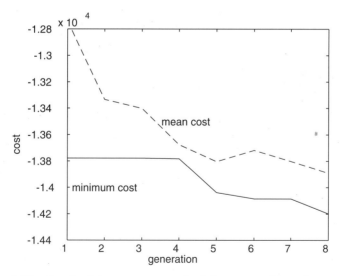

Figure 2.12 Graph of the mean cost and minimum cost for each generation.

2.3   A PARTING LOOK

We've managed to find the highest point in Rocky Mountain National Park with a genetic algorithm. This may have seemed like an easy, trivial problem—the peak could have easily been found through an exhaustive search or by looking at a topographical map. True. But, try using a conventional numerical optimization routine to find the peak in these data. Such routines don't work very well. Many can't even be adapted to apply to this simple problem. We'll present some much more difficult problems in Chapters 4 and 6 where the utility of the genetic algorithm becomes even more apparent. For now you should be comfortable with the workings of a simple genetic algorithm.

To summarize the simple genetic algorithm, we present a pseudocode version in Appendix A. This code lacks a lot of bells and whistles, but should make writing your own genetic algorithm quite easy. The bibliography at the end of this chapter cites very introductory articles that provide an overview of genetic algorithms.

BIBLIOGRAPHY

Angeline, P. J., 1995, "Evolution revolution: An introduction to the special track on genetic and evolutionary programming," *IEEE Expert Intelligent Systems and Their Applications* **10**, June pp. 6–10.

Goldberg, D. E., 1993 Feb., "Making genetic algorithms fly: A lesson from the Wright brothers," *Adv. Technol. Dev.* **2**, pp. 1–8.

Holland, J. H., 1992 July, "Genetic algorithms," *Sci. Am.*, pp. 66–72.

Janikow, C. Z., and D. St. Clair, 1995 Feb./March, "Genetic algorithms simulating nature's methods of evolving the best design solution," *IEEE Potentials* **14**, pp. 31–35.

Malasri, S., J. R. Martin, and L. Y. Lin, 1995 Jan.–March, "Hands-on software for teaching genetic algorithms," *Comput. Educ. J.* **VI**, pp. 42–47.

CHAPTER 3

THE CONTINUOUS PARAMETER GENETIC ALGORITHM

Now that you are convinced (perhaps) that the binary genetic algorithm solves many optimization problems that stump traditional techniques, let's look a bit closer at the quantization limitation. What if you are attempting to solve a problem where the values of the parameters are continuous and you want to know them to the full machine precision? In such a problem, each parameter requires many bits to fully represent it. If the number of parameters is large, the size of the chromosome grows quite quickly. Of course, 1s and 0s are not the only way to represent a parameter. One could, in principle, use any representation conceivable for encoding the parameters. When the parameters are naturally quantized, the binary genetic algorithm fits nicely. However, when the parameters are continuous, it is more logical to represent them by floating-point numbers. In addition, since the binary genetic algorithm has its precision limited by the binary representation of parameters, using real numbers instead easily allows representation to the machine precision. This continuous parameter genetic algorithm also has the advantage of requiring less storage than the binary genetic algorithm because a single floating-point number represents the parameter instead of N_{bits} integers. As N_{bits} increases, this storage becomes significant. Of course, the other advantage is in the accurate representation of the continuous parameter. It follows that the representation of the cost function is also more accurate as a result.

The purpose of this chapter is to introduce the continuous parameter genetic algorithm. Most sources call this version of the genetic algorithm a real-valued genetic algorithm. We use the term continuous parameter rather than real-valued to avoid confusion between real and complex numbers. The development here closely parallels the last chapter. We primarily dwell upon the differences in the two algorithms. The continuous parameter example introduced in Chapter 1 is our primary example problem. This allows the reader to compare the continuous genetic algorithm performance with more traditional optimization algorithms introduced in Chapter 1.

3.1 COMPONENTS OF A CONTINUOUS PARAMETER GENETIC ALGORITHM

The flow chart in Figure 3.1 provides a "big picture" overview of a continuous genetic algorithm. Each block is discussed in detail in this chapter. This genetic algorithm is very similar to the binary genetic algorithm presented in the last chapter. The primary difference is the fact that parameters are no

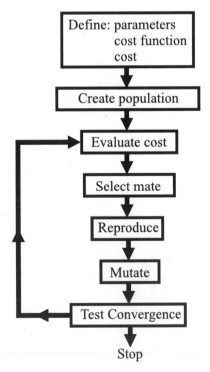

Figure 3.1   Flow chart of a continuous genetic algorithm.

longer represented by bits of zeros and ones, but instead by real numbers over whatever range is deemed appropriate. However, this simple fact adds some nuances to the application of the technique that must be carefully considered. In particular, we will carefully reconsider how to apply our crossover and mutation operators.

3.1.1 The Example Parameters and Cost Function

As we saw in the last chapter, the goal is to solve some optimization problem where we search for an optimal (minimum) solution in terms of the parameters of the problem. Therefore, we begin the process of fitting it to a genetic algorithm by defining a chromosome as an array of parameter values to be optimized. If the chromosome has N_{par} parameters (an N-dimensional optimization problem) given by $p_1, p_2, \ldots, p_{Npar}$ then the chromosome is written as an array with $1 \times N_{par}$ elements so that

$$chromosome = [p_1, p_2, p_3, \ldots, p_{Npar}] \tag{3.1}$$

In this case, the parameters are each represented as floating-point numbers. Each chromosome has a cost found by evaluating the cost function, f, at the parameters $p_1, p_2, \ldots, p_{Npar}$.

$$c = f(chromosome) = f(p_1, p_2, \ldots, p_{Npar}) \tag{3.2}$$

Equations (3.1) and (3.2) along with applicable constraints constitute the problem to be solved.

Our primary example in this chapter is the continuous function introduced in Chapter 1. Consider the cost function

$$c = f(x, y) = x \sin(4x) + 1.1y \sin(2y) \tag{3.3}$$

Subject to the constraint: $0 \leq x \leq 10$ and $0 \leq y \leq 10$

$$\tag{3.4}$$

Since f is a function of x and y only, the clear choice for the parameters is

$$chromosome = [x, y] \tag{3.5}$$

with $N_{par} = 2$. A contour map of the cost function appears as Figure 1.4. This cost function is considerably more complex than the cost function in Chapter 2. We see that peaks and valleys dot the landscape of the cost function contour plot. Traditional minimum seeking methods are overwhelmed by the plethora of local minima. Our goal is to find the global minimum value of $f(x, y)$.

3.1.2   Parameter Encoding, Accuracy, and Bounds

Here is where we begin to see the differences from the prior chapter. We no longer need to consider how many bits are necessary to accurately encode a parameter. Instead, the x and y values are encoded in terms of real numbers that fall between the bounds listed in equation (3.4). Although continuous parameters may assume any value, in reality, a digital computer represents numbers by a finite number of bits. When we refer to the continuous parameter genetic algorithm we mean the computer uses its internal precision and roundoff to define the accuracy of the continuous parameter. Now the algorithm is limited in accuracy to the roundoff error of the computer. Single-precision arithmetic uses 16 bit precision, while double-precision arithmetic uses 32 bit precision on a PC. A supercomputer often uses 64 bit precision.

Since the genetic algorithm is a search technique, it must be limited to exploring a reasonable region of parameter space. Sometimes this is done by imposing a constraint on the problem such as equation (3.4). If one does not know the initial search region, there must be enough diversity in the initial population to explore a reasonable-sized parameter space before focusing on the most promising regions.

3.1.3   Initial Population

To begin the genetic algorithm, we define an initial population of N_{ipop} chromosomes. A matrix represents the population with each row in the matrix being a $1 \times N_{par}$ array (chromosome) of continuous parameter values. Given an initial population of N_{ipop} chromosomes, the full matrix of $N_{ipop} \times N_{par}$ random values is generated by

$$IPOP = (hi - lo) \times \textbf{random}\{N_{ipop}, N_{par}\} + lo$$

where

$\textbf{random}\{N_{ipop}, N_{par}\}$ = a function that generates an $N_{ipop} \times N_{par}$ matrix of uniform random numbers between zero and one

hi = highest number in the parameter range

lo = lowest number in the parameter range

Note that this expression can easily be generalized to allow different values of *lo* and *hi* for each parameter.

This society of chromosomes is not a democracy: the individual chromosomes are not all created equal. Each one's worth is assessed by the cost function. So at this point, the parameters are passed to the cost function for evaluation.

For our example problem equations (3.3), (3.4), and (3.5), we have two parameters ($N_{par} = 2$) and the *x* and *y* values are each constrained to lie between $lo = 0$ and $hi = 10$. This cost function (3.3) is more complex than the one in Chapter 2, so we start this genetic algorithm with a larger initial population. There is no set sampling criterion akin to the Nyquist rate in Fourier analysis; however, a highly undulating cost surface should have more samples than a smooth cost surface. Here we choose $N_{ipop} = 48$ so that the dimension of the initial population matrix is 48×2. The larger population allows the algorithm to sample the cost surface in more detail. There is always a trade-off between the number of generations the algorithm needs to converge versus the size of the initial population. Figure 3.2 shows the initial population for the $N_{ipop} = 48$ random chromosomes. We see widely scattered population members that well sample the values of the cost function (the contours of the plot).

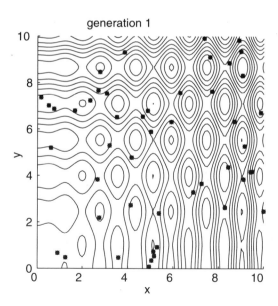

Figure 3.2   Contour plot of the cost function with the initial population for the $N_{ipop} = 24$ random chromosomes indicated by asterisks.

3.1.4 Natural Selection

Now is the time to decide which chromosomes in the initial population are fit enough to survive and possibly reproduce offspring in the next generation. As done for the binary version of the algorithm, the N_{ipop} costs and associated chromosomes are ranked from lowest cost to highest cost. We only retain the best N_{pop} members of the population for the next iteration of the algorithm. The rest die off. This process of natural selection must occur at each iteration of the algorithm to allow the population of chromosomes to evolve over the generations to the most fit members as defined by the cost function. From this point on, the size of the population at each generation is N_{pop}.

Not all of the survivors are deemed fit enough to mate. Of the N_{pop} chromosomes in a given generation, only the top N_{good} are kept for mating and the bottom N_{bad} are discarded to make room for the new offspring (where $N_{good} + N_{bad} = N_{pop}$).

In our example, the mean of the cost function for the initial population of 48 was 0.9039 and the best cost was -16.26. The mean of the population after discarding the bottom half is -4.27. We will use $N_{pop} = 24$ for the population of each iteration of the genetic algorithm. Only the top $N_{good} = 12$ chromosomes survive to mate while the remaining $N_{bad} = 12$ die off. The natural selection results represented to four significant digits are shown in Table 3.1.

3.1.5 Pairing

The $N_{good} = 24$ most fit chromosomes form the mating pool. Six mothers and fathers pair in some random fashion. Each pair produces two offspring that contain traits from each parent. In addition, the two parents survive to be part of the next generation. The more similar the two parents, the more likely are the offspring to carry the traits of the parents. There are various reasonable ways to pair the chromosomes. We presented three basic approaches in Chapter 2 and refer the reader back to that presentation rather than repeating it.

The example presented here uses a weighted cost selection with the probabilities shown in Table 3.2. Parents are rarely selected from the bottom half of the population, because their costs are so much smaller than the parents at the top of the list. A random number generator produced the following six pairs of random numbers: (0.4679, 0.5344), (0.2872, 0.4985), (0.1783, 0.9554), (0.1537, 0.7483), (0.5717, 0.5546), and (0.8024, 0.8907). Using these random pairs and Table 3.2, the following chromosomes were

TABLE 3.1   Population of 24 Chromosomes after Ranking of the Initial Population of 48 Chromosomes Discarding the Bottom Half

Number	x	y	Cost
1	9.0465	8.3097	-16.2555
2	9.1382	5.2693	-13.5290
3	7.6151	9.1032	-12.2231
4	2.7708	8.4617	-11.4863
5	8.9766	9.3469	-10.3505
6	5.9111	6.3163	-5.4305
7	4.1208	2.7271	-5.0958
8	2.7491	2.1896	-5.0251
9	3.1903	5.2970	-4.7452
10	9.0921	3.8350	-4.6841
11	0.6056	5.1942	-4.2932
12	4.1539	4.7773	-3.9545
13	8.4598	8.8471	-3.3370
14	7.2541	3.6534	-1.4709
15	3.8414	9.3044	-1.1517
16	8.6825	6.3264	-0.8886
17	1.2537	0.4746	-0.7724
18	7.7020	7.6220	-0.6458
19	5.3730	2.3777	-0.0419
20	5.0071	5.8898	0.0394
21	0.9073	0.6684	0.2900
22	8.8857	9.8255	0.3581
23	2.6932	7.6649	0.4857
24	2.6614	3.8342	1.6448

TABLE 3.2   Chromosomes in the Mating Pool and Their Associated Cost Weighted Probabilities

n	P_n	$\sum_{i=1}^{n} P_i$
1	0.2265	0.2265
2	0.1787	0.4052
3	0.1558	0.5611
4	0.1429	0.7040
5	0.1230	0.8269
6	0.0367	0.8637
7	0.0308	0.8945
8	0.0296	0.9241
9	0.0247	0.9488
10	0.0236	0.9724
11	0.0168	0.9892
12	0.0108	1.000

randomly selected to mate

$$mom = [3\ 2\ 1\ 1\ 4\ 5]$$

$$dad = [3\ 3\ 10\ 5\ 3\ 7]$$

Thus, $chromosome_3$ mates with $chromosome_3$, and so forth. The mom and dad vectors contain the numbers corresponding to the chromosomes selected for mating.

3.1.6   Mating

As for the binary algorithm, two parents are chosen, and the offspring are some combination of these parents. Many different approaches have been tried for crossing over in continuous parameter genetic algorithms. Adewuya (1996) reviews some of the current methods thoroughly. Several interesting methods are demonstrated by Michalewicz (1994).

The simplest methods choose one or more points in the chromosome to mark as the crossover points. Then the parameters between these points are merely swapped between the two parents. For example purposes, consider the two parents to be

$$parent_1 = [p_{m1}, p_{m2}, p_{m3}, p_{m4}, p_{m5}, p_{m6}, \ldots, p_{mN_{par}}]$$

$$parent_2 = [p_{d1}, p_{d2}, p_{d3}, p_{d4}, p_{d5}, p_{d6}, \ldots, p_{dN_{par}}]$$

Crossover points are randomly selected, then the parameters in between are exchanged:

$$offspring_1 = [p_{m1}, p_{m2}, {}_\uparrow p_{d3}, p_{d4} {}_\uparrow, p_{m5}, p_{m6}, \ldots, p_{mN_{par}}]$$

$$offspring_2 = [p_{d1}, p_{d2}, {}_\uparrow p_{m3}, p_{m4} {}_\uparrow, p_{d5}, p_{d6}, \ldots, p_{dN_{par}}]$$

The extreme case is selecting N_{par} points and randomly choosing which of the two parents will contribute its parameter at each position. Thus, one goes down the line of the chromosomes and at each parameter, randomly chooses whether or not to swap information between the two parents. This method is called uniform crossover.

$$offspring_1 = [p_{m1}, p_{d2}, p_{d3}, p_{d4}, p_{d5}, p_{m6}, \ldots, p_{dN_{par}}]$$

$$offspring_2 = [p_{d1}, p_{m2}, p_{m3}, p_{m4}, p_{m5}, p_{d6}, \ldots, p_{mN_{par}}]$$

The problem with these point crossover methods is that no new information is introduced: each continuous parameter value that was randomly initiated in the initial population is propagated to the next generation, only in different combinations. Although this strategy worked fine for binary representations, now there is a continuum of values of which we are merely interchanging two data points.

The blending methods remedy this problem by finding ways to combine parameter values from the two parents into new parameter values in the offspring. A single offspring parameter value, p_{new}, comes from a combination of the two corresponding offspring parameter values (Radcliff, 1991)

$$p_{new} = \beta p_{mn} + (1 - \beta)p_{dn} \qquad (3.6)$$

where

β = random number on the interval [0, 1]
p_{mn} = the nth parameter in the mother chromosome
p_{dn} = the nth parameter in the father chromosome

The same parameter of the second offspring is merely the complement of the first (i.e., replacing β by $1 - \beta$). If $\beta = 1$, then p_{mn} propagates in its

entirety and p_{dn} dies. In contrast, if $\beta = 0$, then p_{dn} propagates in its entirety
and p_{mn} dies. When $\beta = 0.5$ (Davis, 1991), the result is an average of the
parameters of the two parents. This method is demonstrated to work well
on several interesting problems by Michalewicz (1994). Choosing which
parameters to blend is the next issue. Sometimes, this linear combination
process is done for all parameters to the right or to the left of some crossover
point. Any number of points can be chosen to blend, up to N_{par} values
where all parameters are linear combinations of those of the two parents.
The parameters can be blended by using the same β for each parameter
or by choosing different β's for each parameter. These blending methods
effectively combine the information from the two parents and choose values
of the parameters between the values bracketed by the parents; however,
they do not allow introduction of values beyond the extremes already
represented in the population. To do this requires an extrapolating method.
The simplest of these methods is linear crossover (Wright, 1991). In this
case, three offspring are generated from the two parents by

$$p_{new1} = 0.5p_{mn} + 0.5p_{dn}$$

$$p_{new2} = 1.5p_{mn} - 0.5p_{dn}$$

$$p_{new3} = -0.5p_{mn} + 1.5p_{dn}$$

Any parameter outside the bounds is discarded in favor of the other two.
Then the best two offspring are chosen to propagate. Of course, the factor
0.5 is not the only one that can be used in such a method. Heuristic
crossover (Michalewicz, 1991) is a variation where some random number,
β, is chosen on the interval $[0, 1]$ and the parameters of the offspring are
defined by

$$p_{new} = \beta(p_{mn} - p_{dn}) + p_{mn}$$

Variations on this theme include choosing any number of parameters
to modify and generating different β for each parameter. This method
also allows generation of offspring outside of the values of the two parent
parameters. Sometimes values are generated outside of the allowed range.
If this happens the offspring is discarded and the algorithm tries another β.
The blend crossover (BLX-α) method (Eshelman and Shaffer, 1993) begins
by choosing some parameter α that determines the distance outside the
bounds of the two parent parameters that the offspring parameter may lie.
This method allows new values outside of the range of the parents without
letting the algorithm stray too far. Many codes combine the various methods
to use the strengths of each. New methods, such as quadratic crossover

(Adewuya, 1996), do a numerical fit to the fitness function. Three parents are necessary to perform a quadratic fit.

The algorithm used most often in this book is a combination of an extrapolation method with a crossover method. We wanted to find a way to closely mimic the advantages of the binary genetic algorithm mating scheme. It begins by randomly selecting a parameter in the first pair of parents to be the crossover point

$$\alpha = \textbf{roundup}\{\textbf{random} * N_{par}\}$$

We'll let

$$parent_1 = [p_{m1}p_{m2} \cdots p_{m\alpha} \cdots p_{mN_{par}}]$$

$$parent_2 = [p_{d1}p_{d2} \cdots p_{d\alpha} \cdots p_{dN_{par}}]$$

where the m and d subscripts discriminate between the *mom* and the *dad* parent. Then the selected parameters are combined to form new parameters that will appear in the children:

$$p_{new1} = p_{m\alpha} - \beta[p_{m\alpha} - p_{d\alpha}]$$

$$p_{new2} = p_{d\alpha} + \beta[p_{m\alpha} - p_{d\alpha}]$$

where β is also a random value between 0 and 1. The final step is to complete the crossover with the rest of the chromosome as before:

$$offspring_1 = [p_{m1}p_{m2} \cdots p_{new1} \cdots p_{dN_{par}}]$$

$$offspring_2 = [p_{d1}p_{d2} \cdots p_{new2} \cdots p_{mN_{par}}]$$

If the first parameter of the chromosomes is selected, then only the parameters to the right of the selected parameter are swapped. If the last parameter of the chromosomes is selected, then only the parameters to the left of the selected parameter are swapped. This method does not allow offspring parameters outside the bounds set by the parent unless $\beta > 1$.

For our example problem, the first set of parents are identical and produce clones of themselves. Rather uninteresting, so let's move on to the second pair. *chromosome*$_2$ and *chromosome*$_3$ are given by

$$chromosome_2 = [5.2693, 9.1382]$$

$$chromosome_3 = [9.1032, 7.6151]$$

A random number generator selects p_1 as the location of the crossover. A second random number for β is selected: $\beta = 0.7147$. The new offspring are given by

$$offspring_3 = [5.2693 - 0.7147 \times 5.2693 + 0.7147 \times 9.1032, \ 7.6151]$$

$$= [8.0094, \ 7.6151]$$

$$offspring_4 = [9.1032 + 0.7147 \times 5.2693 - 0.7147 \times 9.1032, \ 9.1382]$$

$$= [6.3631, \ 9.1382]$$

Continuing this process with the other four sets of parents produces eight more offspring.

3.1.7 Mutations

Here, as in the last chapter, we can sometimes find our method working a bit too well. If care is not taken, the genetic algorithm converges too quickly into one region of the cost surface. If this area is in the region of the global minimum, that is good. However, some functions, such as the one we are modeling, have many local minima. If we do nothing to solve this tendency to converge quickly, we could end up in a local rather than a global minimum. To avoid this problem of overly fast convergence, we force the routine to explore other areas of the cost surface by randomly introducing changes, or mutations, in some of the parameters. For the binary algorithm, this amounted to just changing a bit from a 0 to a 1 or vice versa. The basic method of mutation is not much more complicated for the continuous parameter genetic algorithm (although more complicated methods exist [Michalewicz, 1994]).

As with the binary genetic algorithm, a mutation rate between 1 and 20% often works well. If the mutation rate is above 20%, too many good parameters are mutated, and the algorithm stalls. Multiplying the mutation rate by the total number of parameters gives the number of parameters that should be mutated. Next, random numbers are chosen to select the row and columns of the parameters to be mutated. A mutated parameter is replaced by a new random parameter.

We applied a mutations at a rate of 4% ($\mu = 0.04$) for our example problem. Since there are 12 chromosomes before mating and 24 afterwards, this amounted to a total of 2 mutations ($0.04 \times 24 \times 2$). The following parameters are mutated: p_1 of $chromosome_7$ and p_2 of $chromosome_{22}$. Thus, p_1 of $chromosome_7$ is deleted and replaced with a new uniform random number between 0 and 10.

$$chromosome_7 = [2.7271, \ 4.1208]$$
$$\Downarrow$$
$$chromosome_7 = [3.1754, \ 4.1208]$$

After the new offspring are born and the mutations induced, Table 3.3 shows the second generation ranked from best to worst. The mean for this population is -8.51.

TABLE 3.3   Second Generation of Continuous Parameters

x	y	Cost
9.0465	8.3128	-16.2929
9.0465	8.3097	-16.2555
9.1382	5.2693	-13.5290
7.6151	9.1032	-12.2231
7.6151	9.1032	-12.2231
7.6151	9.1032	-12.2231
2.7708	8.4789	-11.6107
2.7708	8.4617	-11.4863
9.1382	8.0094	-11.0227
8.9766	9.3438	-10.4131
8.9766	9.3469	-10.3505
9.0465	7.9025	-9.8737
1.5034	9.0860	-6.6667
4.4224	9.3469	-5.6490
5.9111	6.3163	-5.4305
7.6151	6.3631	-5.1044
9.0921	4.2422	-5.0619
2.7491	2.1896	-5.0251
3.1903	5.2970	-4.7452
9.0921	3.8350	-4.6841
0.6056	5.1942	-4.2932
4.1539	4.7773	-3.9545
8.6750	2.7271	-3.4437
4.1208	3.1754	-2.6482

3.1.8 Convergence

This run of the algorithm found the minimum cost (-18.5) in seven generations. Members of the population are shown as asterisks on the cost surface contour plot in Figures 3.3 to 3.5 for generations 3, 5, and 7, respectively. The members of the population begin gathering around two minima by generation 3. By generation 5, they target these two minima and get closer to the true bottoms. The global minimum of -18.5. is found in generation 7. All but two of the population members are in the valley of the global minimum in the final generation. Figure 3.6 is a plot of the mean and minimum cost for each generation. The genetic algorithm was able to find the global minimum, unlike the Nelder-Mead and BFGS algorithms presented in Chapter 1.

3.2 A PARTING LOOK

The binary genetic algorithm could have been used in this example as well as a continuous parameter genetic algorithm. Since the problem used continuous parameters, it seemed more reasonable to use the continuous

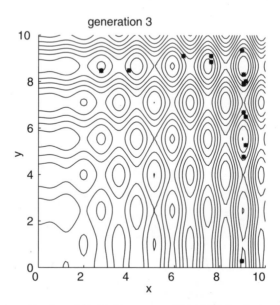

Figure 3.3 Contour plot of the cost function with the population at generation 3 ($N_{pop} = 24$) indicated by asterisks.

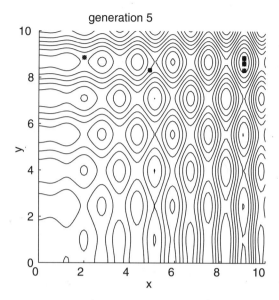

Figure 3.4   Contour plot of the cost function with the population at generation 5 (N_{pop} = 24) indicated by asterisks.

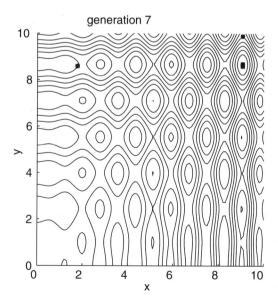

Figure 3.5   Contour plot of the cost function with the population at generation 7 (N_{pop} = 24) indicated by asterisks.

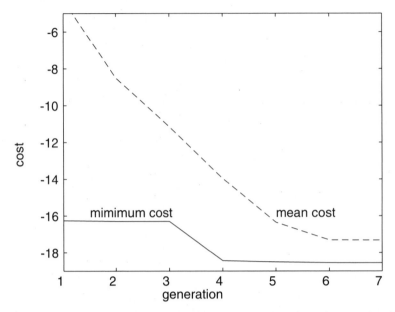

Figure 3.6   Plot of the minimum and mean costs as a function of generation. The algorithm converged in seven generations.

parameter genetic algorithm. The next chapter presents some practical optimization problems for both the binary and continuous parameter genetic algorithms. Selecting the various genetic algorithm parameters, such as mutation rate and type of crossover, is still more of an art than a science.

BIBLIOGRAPHY

Adewuya, A. A., 1996, *New Methods in Genetic Search with Real-Valued Chromosomes*, Master's Thesis, Cambridge: Massachusetts Institute of Technology.

Davis, L., 1991, "Hybridization and numerical representation," in L. Davis (Ed.), *The Handbook of Genetic Algorithms*, New York: Van Nostrand Reinhold, pp. 61–71.

Eshelman, L. J., and D. J. Shaffer, 1993, "Real-coded genetic algorithms and interval-schemata," in D. L. Whitley (Ed.), *Foundations of Genetic Algorithms 2*, San Mateo, CA: Morgan Kaufman, pp. 187–202.

Michalewicz, Z., 1994, *Genetic Algorithms + Data Structures = Evolution Programs*, 2nd ed., New York: Springer-Verlag.

Radcliff, N. J., 1991, "Forma analysis and random respectful recombination," in *Proc. of the Fourth International Conference on Genetic Algorithms*, San Mateo, CA: Morgan Kauffman.

Wright, A., 1991, "Genetic algorithms for real parameter optimization," in G. J. E. Rawlins (Ed.), *Foundations of Genetic Algorithms*, San Mateo, CA: Morgan Kaufmann, pp. 205–218.

CHAPTER 4

APPLICATIONS

The examples in this chapter make use of the genetic algorithms introduced in Chapters 2 and 3. Enough details are provided for the user to try the problems too. Remember, the genetic algorithm uses a random number generator, so the exact results won't be reproducible. The first three examples are somewhat generic, using methods directly applicable to a wide range of problems. The last one is arcane in that it deals with an extremely problem-specific cost function. However, the solution method is very straightforward, and the subroutine for the cost function could be replaced to solve many problems in the same vein.

4.1 "MARY HAD A LITTLE LAMB"

In the first example, we'll test the musical talent of the genetic algorithm to see if it can learn the first four measures of "Mary Had a Little Lamb." This song is in the key of C with 4/4 time and only has quarter and half notes. In addition, the frequency variation of the notes is less than an octave. A chromosome for this problem has $4 \times 4 = 16$ genes (one gene for each beat). The binary genetic algorithm is perfect, because there are eight distinct notes, seven possible frequencies, and one hold. (The encoding is given in Table 4.1.) A hold indicates the previous note is a

TABLE 4.1 Binary Encoding of the Musical Notes for "Mary Had a Little Lamb"

Code	note
000	hold
001	A
010	B
011	C
100	D
101	E
110	F
111	G

half note. (It is possible to devote one bit to indicate a quarter or half note. However, this approach would require variable-length chromosomes.) The actual solution only makes use of the C, D, E, and G notes and the hold. Therefore, the binary encoding is only 62.5% efficient. Since the exact notes in the song may not be known, the inefficiency is only evident in hindsight. An exhaustive search would have to try $8^{16} = 2.8147 \times 10^{14}$ possible combinations of notes.

In this case, we know the correct answer, and the notes are shown in Figure 4.1. Putting the song parameters into a row vector yields

$$[EDCDEEEholdDDDholdEGGhold]$$

with a corresponding chromosome encoded as

ans
$$= [101\,100\,011\,100\,101\,101\,101\,000\,100\,100\,100\,000\,101\,111\,111\,000]$$

We'll attempt to find this solution two ways. First, the computer compares a chromosome with the known answer and gives a point for each bit in the proper location. Second, we'll use our ear (very subjective) to rank the chromosomes.

Figure 4.1 Music to "Mary Had a Little Lamb."

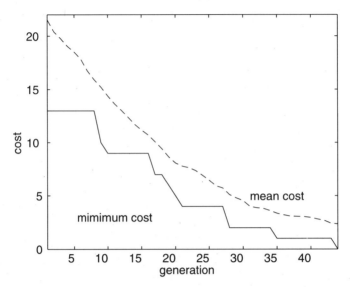

Figure 4.2   The minimum cost and mean cost as a function of generation when the computer knows the exact answer (all the notes to "Mary Had a Little Lamb").

The first cost function subtracts the binary guess (*guess*) from the true answer (*answer*) and sums the absolute value of all the digits.

$$cost = \sum_{n=1}^{48} |answer[n] - guess[n]| \tag{4.1}$$

When a chromosome and the true answer match, the cost is zero. The genetic algorithm begins with an initial population of 96 members and $N_{pop} = 48$, *keep* $= 24$, and $\mu = 0.05$. Figure 4.2 shows an example of the cost function statistics as a function of generation. The genetic algorithm consistently finds the correct answer over many different runs.

The second cost function is subjective. This is an interesting twist in that it combines the computer with a human response to judge performance. Cost is assigned from a 0 (that's the song) to a 100 (terrible match). This algorithm wears on your nerves, since some collection of notes must be played and judged for each chromosome. Consequently, the authors could only listen to the first two measures and judge the performance. Figure 4.3 shows the cost function statistics as a function of generation. Compare this graph with the previous one.

Subjective cost functions are quite fascinating. A chromosome in the first generations tended to receive a lower cost than the identical chromo-

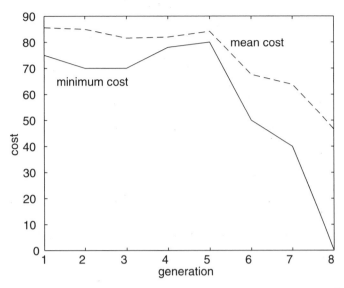

Figure 4.3 The minimum cost and mean cost as a function of generation when the subjective cost function was used (our judgement as to how close the song associated with a chromosome came to the actual tune of "Mary Had a Little Lamb").

some in later generations. This accounts for the increase in the minimum cost shown in Figure 4.3. Many chromosomes produce very unmelodic sounds. Changing expectations or standards is common in everyday life. For instance, someone who begins a regular running program is probably happier with a 10 minute mile than she would be after one year of training. The subjective cost function converged faster than the mathematical cost function, because a human is able to evaluate more aspects of the experiment than the computer. Consider the difference between a G note and the C, E, and F notes. The human ear ranks the three notes according to how close they are to G, while the computer only notices that the three notes differ from G by one bit. A continuous parameter genetic algorithm might work better in this situation.

Extensions to this idea are limitless. Composing music (Biles, 1994; Jacob, 1995) and producing art with a genetic algorithm are interesting possibilities. Putnam (1994) tried his hand at composing music with a genetic algorithm. The idea was to construct random pieces of music and rate them for their musical value. Putnam used two approaches. The first combined mathematical functions (like sine and cosine) in the time domain. This approach produced too much noise, and people had difficulty judging the performance (patience and tolerance were difficult). The second approach used notes as in our example. This approach worked better, but listening to many randomly generated music pieces is too difficult for most of us. He concluded:

> The primary problems with these programs seem to be those involving the human interface. The time it takes to run the program and listen to the samples combined with the relative dullness of the samples mean that the users rapidly get tired and bored. This means that the programs need to be built to generate the samples relatively quickly, the samples need to be relatively short, the samples need to develop and evolve to be interesting quickly and of necessity the population size must be small.

A few years ago fractal and chaotic art were very popular items. Maybe genetic algorithm art is in our future. Fractal movies have been made using genetic programming (Angeline, 1996). These movies were generated with some input by the user. Other researchers have color mapped genetic algorithm convergence to a phase space to create beautiful fractal images (Juliany and Vose, 1993). These images resemble the fractal art generated by applying iterative methods to solving nonlinear equations.

The idea of a subjective cost function is closely tied to work with fuzzy sets and logic (Ross, 1995). In classical set theory, an element is either a member or not. In fuzzy set theory, an element can have various degrees of membership. Thus, the boundaries between fuzzy sets are vague and ambiguous. Is this piece of art or music good? We're likely to be fuzzy about our decision and assign a cost between zero and one rather than either a zero or a one.

Most companies use computers to design products. Various parameters, bounds, and tolerances are input to the computer and eventually a design pops out. Perhaps a better approach is to have a closer interaction between the computer and the designer. A computer calculates well and does mundane tasks such as sorting a list well, but it is not good at making subjective judgements. If the computer program occasionally stops and presents some options and their associated costs to the designer, then the designer has the ability to add a human element to the design. Since the genetic algorithm carries a list of answers in descending order of preference, it is perfect for fuzzy design applications.

4.2 WORD GUESS

In this game, the genetic algorithm is given the number of letters in a word, and it guesses the letters that compose the word until it finds the right answer. In this case, we'll use a genetic algorithm where each letter is given the integer corresponding to its location in the alphabet (e.g., a = 1, b = 2, etc.). Establishing the rules of the game determines the shape of the cost surface. If the cost is the sum of the squares of the differences of the numbers representing the letters in the chromosome (computer's guess at the word) and the true answer, then the surface is described by

$$cost = \sum_{n=1}^{\text{# letters}} (guess[n] - answer[n])^2 \qquad (4.2)$$

where

$\text{\# letters}$ = number of letters in the word
$guess[n]$ = letter n in the guess chromosome
$answer[n]$ = letter n in the answer

Figure 4.4 shows the cost surface for the two-letter word "he" given the cost function in equation (4.2). There are a total of $26^2 = 676$ possible combinations to check. For N letters, there are 26^N possible combinations. The unknown word that the genetic algorithm must find is "colorado." The genetic algorithm begins with $N_{ipop} = 64$, $N_{pop} = 32$, $keep = 16$, and $\mu = 0.04$. After 27 iterations, the genetic algorithm correctly guesses the word. The guesses as a function of generation are given in Table 4.2.

A slight change to the previous cost function creates a completely different cost surface. This time, a correct letter is given a zero, while an

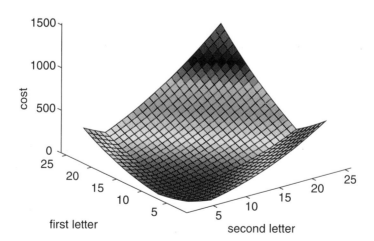

Figure 4.4   The cost surface for the two-letter word "he" associated with Equation (4.2). There a total of $26^2 = 676$ possible combinations.

**TABLE 4.2   Genetic Algorithm's Best Guess
(First Cost Function) after Each Generation**

Generation	Best Guess
1	dhgmtfbn
2	bmloshjm
3	bmloshjm
4	bmlorfds
5	bmlorfds
6	bmlorddm
7	bmlorddm
8	bmlorddm
9	bmlosadn
10	bmlosadn
11	bmlosadn
12	bmlosadn
13	bmlosadn
14	cmlorbdo
15	cmlorbdo
16	cmlorbdo
17	cmlorbdo
18	cmlorbdo
19	cmlorbdo
20	colorbdo
21	colorbdo
22	colorbdo
23	colorbdo
24	colorbdo
25	colorbdo
26	colorbdo
27	colorado

incorrect letter is given a one. There is no gray area associated with this cost function.

$$cost = \sum_{n=1}^{\# \, letters} 1 - sgn(guess_n - answer_n) \qquad (4.3)$$

where

$$sgn(x) = \begin{cases} 1 \text{ when } x = 0 \\ 0 \text{ when } x \neq 0 \end{cases} \qquad (4.4)$$

Figure 4.5 shows the cost surface for the two-letter word "he" with the cost function given by equation (4.3). We now apply this function to the word "colorado." After 17 generations and $N_{ipop} = 64$, $N_{pop} = 32$, $keep = 16$, and $\mu = 0.04$, the genetic algorithm correctly guesses the word (colorado) as shown in Table 4.3. This is quite an accomplishment given that the total number of possible combinations is $26^8 = 2.0883 \times 10^{11}$. Part of the reason for the success of the second cost function is that the second cost function has a large cost differential between a correct and incorrect letter, while the first cost function assigns a cost depending upon the proximity of the correct and incorrect letters. As a result, a chromosome with all wrong letters, but whose letters are close to the correct letters, receives a low cost from the second cost function, but a high cost from the first cost function.

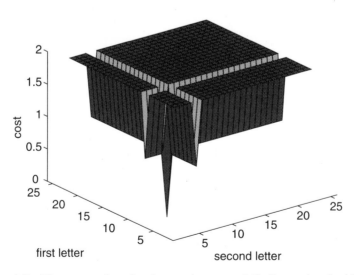

Figure 4.5 The cost surface for the two-letter word "he" associated with equation (4.3).

TABLE 4.3   Genetic Algorithm's Best Guess (Second Cost Function) after Each Generation

Generation	Best Guess
1	pxsowqdo
2	pxsowqdo
3	pxsowqdo
4	bodokado
5	bodokado
6	bodokado
7	bodokado
8	dslorado
9	dslorado
10	dslorado
11	dslorado
12	dslorado
13	cozorado
14	cozorado
15	cozorado
16	cozorado
17	colorado

When cost weighting determines the mating pool, the second cost function tends to have parents with more correct letters than the first cost function would.

4.3   LOCATING AN EMERGENCY RESPONSE UNIT

An emergency response unit is to be built that will best serve a city. The goal is to provide the minimum response time to a medical emergency that could occur anywhere in the city. After a survey of past emergencies, a map is constructed showing the frequency of an emergency in a given section of the city. This example was inspired by an example from a book by Meerschaert (1993) in which the location of a fire station in a community broken into a 6 × 6 square grid and with a cost function similar to equation (4.5) is to be found. The author used a random search algorithm instead of a gradient-based method, because even though $\nabla cost$ can be calculated, it is not possible to algebraically solve $\nabla cost = 0$. This problem seemed too easy for the genetic algorithm, so we increased the grid size and added the constraints. The city is divided into a grid of 10 km $\times$ 10 km with 100 sections, as shown in Figure 4.6. The response time of the fire station is estimated to be $1.7 + 3.4r$ minutes, where r is in kilometers. This formula is not based on real data, but an actual city would have an estimate of this formula based on traffic, time of day, and so on. An appropriate cost function is the sum of the distances weighted by the frequency of emergencies or

$$cost = \sum_{n=1}^{100} w_n \sqrt{\left(x_n - x_{fs}\right)^2 + \left(y_n - y_{fs}\right)^2} \qquad (4.5)$$

where

(x_n, y_n) = coordinates of the center of square n
(x_{fs}, y_{fs}) = coordinates of the proposed emergency response unit
$\quad w_n$ = fire frequency in square n (as shown in Figure 4.6)

The cost surface for this problem is shown in Figure 4.7. It appears to be a nice bowl-shaped surface that minimum seeking algorithms love. The problem for many algorithms is the discrete weighting assigned to the city squares and the small discontinuities in the cost surface that are not apparent from the graph.

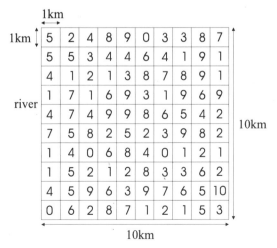

Figure 4.6   A model of a 10 km $\times$ 10 km city divided into 100 equal squares.

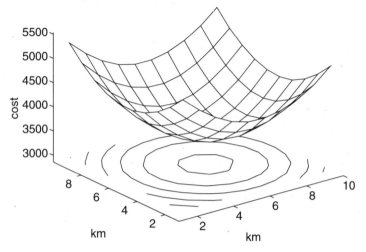

Figure 4.7 Cost surface associated with the cost function in equation (4.5).

On the other hand, including some constraints complicates the cost surface, so it would be difficult for a minimum seeking algorithm to find the bottom. Consider adding a river with only two bridges that cross it. The river is located at $y = 6$ km from the bottom and the bridges cross at $y = 1.5$ and 6.5 km from the left (as shown in Figure 4.8). This new

5	2	4	8	9	0	3	3	8	7
5	5	3	4	4	6	4	1	9	1
4	1	2	1	3	8	7	8	9	1
1	7	1	6	9	3	1	9	6	9
4	7	4	9	9	8	6	5	4	2
7	5	8	2	5	2	3	9	8	2
1	4	0	6	8	4	0	1	2	1
1	5	2	1	2	8	3	3	6	2
4	5	9	6	3	9	7	6	5	10
0	6	2	8	7	1	2	1	5	3

1km

1km

river

10km

y

x

10km

Figure 4.8 A model of a 10 km × 10 km city divided into 100 equal squares with a river and two bridges.

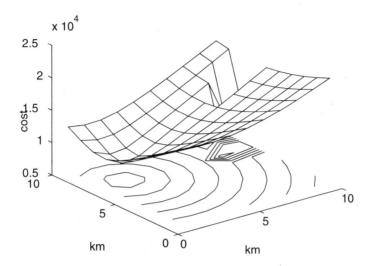

$\times 10^4$

Figure 4.9 Cost surface associated with the cost function in equation (4.5) and the added constraint of a river and two bridges.

cost surface has two distinct minima, as shown in Figure 4.9. The solution found by the genetic algorithms is to place the fire station close to the bridge at $(x, y) = (6.6, 6)$.

We used both a binary genetic algorithm and a continuous parameter genetic algorithm with $N_{ipop} = 40$, $N_{pop} = 20$, $N_{good} = 10$, and $\mu = 0.2$. Each algorithm was run 20 different times with 20 different random seeds. The average minimum cost of a population as a function of generation is shown in Figure 4.10. The continuous parameter genetic algorithm outperformed the binary genetic algorithm by finding a much lower minimum cost over the 25 generations. Ultimately, the binary genetic algorithm only took one more generation to find the minimum than the continuous parameter genetic algorithm.

4.4 ANTENNA ARRAY DESIGN

Satellite communication systems use antennas to receive signals transmitted from a satellite. The antenna has a mainbeam and sidelobes. The mainbeam points into space in the direction of the satellite and has a high gain (gain times received signal power equals power sent to receiver) to amplify the weak signals. Sidelobes have low gains and point in various directions other than the mainbeam. Figure 4.11 shows a typical antenna pattern with a mainbeam and sidelobes. The problem with sidelobes is

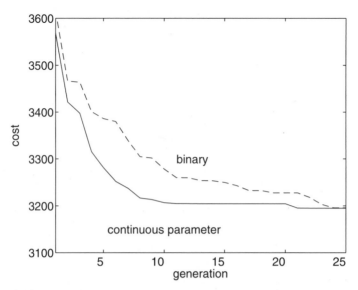

Figure 4.10 Plot of the minimum of the population as a function of generation for the binary genetic and continuous parameter genetic algorithms applied to the emergency response unit problem. These results were averaged over 20 different runs.

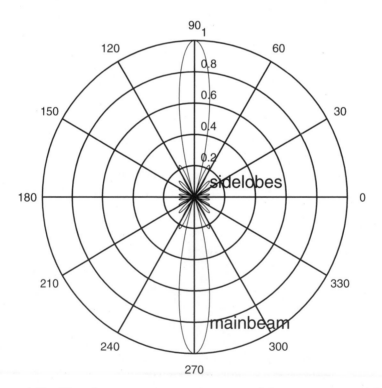

Figure 4.11 Plot of an antenna pattern (response of the antenna vs. angle) that shows the mainbeam and sidelobes.

that strong undesirable signals may enter them and drown out the weaker desired signal entering the mainbeam. Consider a satellite antenna that points its mainbeam in the direction of a satellite. The satellite signal is extremely weak, because it travels a long distance and the satellite transmits a low power. If a cellular phone close to the satellite antenna operates at the same frequency as the satellite signal, the phone signal could enter a sidelobe of the satellite antenna and interfere with its desired signal. Thus, engineers are motivated to maximize the mainbeam gain while minimizing the sidelobe gain.

One type of satellite antenna is the antenna array. A key feature of this antenna is the ability to reduce the gain of the sidelobes. An antenna array is a group of individual antennas that add their signals together to get a single output. The received signals at each of the antenna elements has an amplitude and phase that is a function of frequency, element positions, and angle of incidence of the received signal. The output of the array is a function of the weighting of the signals at the elements. It is possible to weight the amplitudes of the signals at the elements to reduce or eliminate sidelobes. This example shows how to use a genetic algorithm to design a low sidelobe antenna array.

The linear array model has point sources lying along the x-axis (Figure 4.12), and the amplitude taper is symmetric about the center of the array. Its mathematical formulation when the mainbeam points at 90° is given by

$$AF(\phi) = \sum_{n=1}^{N} a_n e^{j(n-1)\Psi} \tag{4.6}$$

where

N = number of elements = $2N_{par}$
$\Psi = kdu = kd \cos \phi$

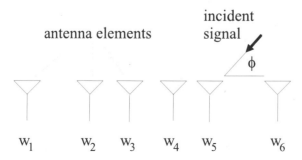

Figure 4.12 Model of a linear array of antenna elements.

a_n=array amplitude weight at element n, $a_m = a_{N+1-m}$
for $m = 1, 2, \ldots, N/2$
$k = \frac{2\pi}{\lambda}$
λ=wavelength
d=spacing between elements
ϕ=angle of incidence of electromagnetic plane wave

The goal is to find the a_n in this formula that yield the lowest possible sidelobe levels in the antenna pattern.

There is a solution to this problem that produces sidelobes that are $-\infty$ dB below the peak of the mainbeam: in other words, no sidelobes at all. The analytical solution is called the binomial array, and the amplitude weights are just the binomial coefficients. Thus, a five-element array has weights that assume the coefficients of a binomial polynomial with five coefficients. Binomial coefficients of an $(N-1)$th-order polynomial, or binomial weights of an N element array, are the coefficients of the polynomial $(z + 1)^{N-1}$ given by the Nth row of Pascal's triangle:

$$
\begin{array}{ccccccc}
 & & & 1 & & & \\
 & & 1 & & 1 & & \\
 & & 1 & 2 & 1 & & \\
 & 1 & 3 & & 3 & 1 & \\
 1 & & 4 & 6 & 4 & & 1 \\
 1 & & & \vdots & & & 1
\end{array}
\qquad
\begin{array}{l}
1 \\
(z + 1) \\
(z + 1)^2 \\
(z + 1)^3 \\
(z + 1)^4 \\
(z + 1)^{N-1}
\end{array}
$$

Figure 4.13 shows a plot of the binomial array antenna pattern. Note the lack of sidelobes.

Our first attempt tried to eliminate the sidelobes of a 42-element array with $d = 0.5\lambda$. Both the binary and continuous parameter genetic algorithms failed to find an amplitude taper that produced maximum sidelobe levels less than -40 dB below the peak of the mainbeam. This result was very disappointing. The problem centers around the cost function formulation. The cost function was the maximum sidelobe level of equation (4.6) with the parameters being a_n. This formulation is difficult to implement and allows undesirable solutions, such as shoulders on the mainbean, that confuse the genetic algorithm.

A different cost function worked much better. The new formulation makes a substitution of variables in equation (4.6)

$$z = e^{j\Psi} \tag{4.7}$$

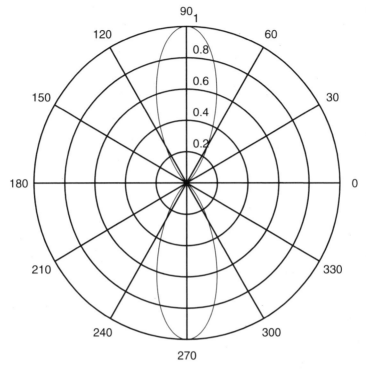

Figure 4.13   Plot of the antenna pattern of an array with binomial weights.

This substitution is known as a z-transform and converts equation (4.6) to

$$AF(\phi) = \sum_{n=1}^{N} a_n z^{n-1} = (z - z_1)(z - z_2)\cdots(z - z_{N-1}) \qquad (4.8)$$

where $z_m = e^{jkdu_m}$. The cost function is the maximum sidelobe level of equation (4.8) with $u_m = \cos\phi_m$ as the parameters. What a difference! As mentioned in Chapter 2, the cost function design is extremely important. Figure 4.14 shows the convergence of the continuous parameter genetic algorithm with $N_{par} = 21$, $N_{ipop} = 32$, $N_{pop} = 16$, $keep = 8$, and $\mu = 0.2$. This excellent performance is exceeded by the binary genetic algorithm with $N_{gene} = 10$, $N_{par} = 21$, $N_{ipop} = 32$, $N_{pop} = 16$, $keep = 8$, and $\mu = 0.2$ (Figure 4.15). Since these algorithms are random, perhaps the binary genetic algorithm won due to chance. To reduce the impact of chance, both algorithms were run ten times with a different random seed each time. They ran for 75 generations with $N_{par} = 21$, $N_{ipop} = 256$,

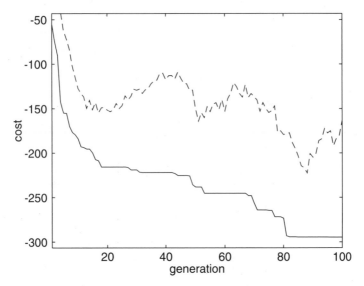

Figure 4.14   Plot of the mean and minimum of the population as a function of generation for the continuous parameter genetic algorithm applied to the antenna design problem.

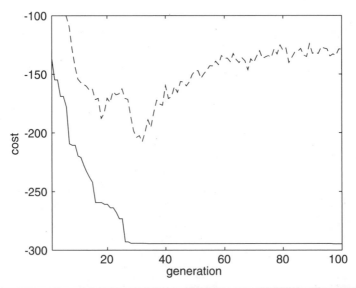

Figure 4.15   Plot of the mean and minimum of the population as a function of generation for the binary genetic algorithm applied to the antenna design problem.

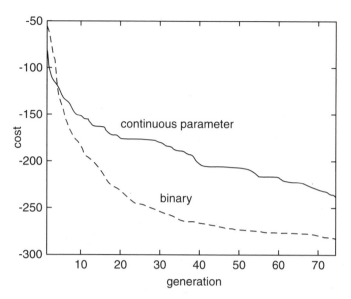

Figure 4.16 Convergence of the binary and continuous parameter genetic algorithms as a function of generation when averaged over ten different runs.

$N_{pop} = 128$, $keep = 64$, and $\mu = 0.2$. Figure 4.16 shows the results of the average minimum cost at each generation. Again, the binary genetic algorithm wins.

Why does the binary genetic algorithm outperform the continuous parameter genetic algorithm? Perhaps it is the size of the search space that must be considered. In this example, the binary genetic algorithm searches over 10^{21} possible solutions. This very large number is small compared to the ∞ number of possible combinations with the continuous parameter genetic algorithm.[1] In any event, the large number of parameters miffs minimum seeking algorithms. For example, the Nelder Mead algorithm failed to converge after 4000 iterations and a random starting point.

4.5 SUMMARY

We've shown a variety of applications of the genetic algorithm. Although the selections were mostly nontechnical, one can imagine technical counterparts. Hopefully, you can think of some interesting applications of the genetic algorithm too. These problems were solved using the simple genetic

---

[1]This is not quite correct, because the computer is limited in accuracy by the number of bits used to represent floating-point numbers.

algorithms introduced in Chapters 2 and 3. The next chapter introduces some analysis of genetic algorithms and some fine points for tuning the algorithms for best performance. Many real world problems have very complex models with cost functions that are time-consuming to evaluate. Finding the optimum population sizes, mutation rates, and other genetic algorithm parameters becomes extremely important.

BIBLIOGRAPHY

Angeline, P. J., 1996, "Evolving Fractal Movies," *Proc. of the First Annual Conference on Genetic Programming*, pp. 503–511.

Biles, J. A., 1994, "GenJam: A genetic algorithm for generating jazz solos," *Proc. of the International Computer Music Conference*.

Jacob, B. L., 1995 Sept., "Composing with genetic algorithms," *Proc. of the International Computer Music Conference*.

Juliany, J., and M. D. Vose, 1993, "The genetic algorithm fractal," *Proc. of the Fifth International Conference on Genetic Algorithms*, p. 639.

Meerschaert, M. M., 1993, *Mathematical Modeling*, Boston: Academic Press, pp. 66–70.

Putnam, J. B., 1994 Aug. 30, "Genetic programming of music," New Mexico Institute of Mining and Technology, Socorro, NM.

Ross, T. J., 1995, *Fuzzy Logic with Engineering Applications*, New York: McGraw-Hill.

CHAPTER 5

AN ADDED LEVEL
OF SOPHISTICATION

Now that we have introduced the genetic algorithm (GA) and discussed in detail the workings of both the binary and the real GA, it is time to look at some of the tricks of the trade. At times, there may be a cost function that is difficult or expensive to evaluate. Other times, we may notice that the usual crossover technique may be destroying some of the best portions of the gene. Deciding on the optimum population size or accuracy needed is not always a trivial task. Since there are various ways to test for convergence, how do we decide which is the best method for our problem? Are there better ways to execute the genetic operators—crossover and mutation? How do we deal with storage issues? Can a genetic algorithm deal with problems that require a specific ordering of the solution? Although there are not always clear-cut solutions to all of these dilemmas, we discuss these issues in this chapter. We give hints and strategies on running a genetic algorithm as well as some useful additions to the codes. The literature is full of other helpful ideas that can be applied to a plethora of problems.

5.1 HANDLING EXPENSIVE COST FUNCTIONS

Sometimes the cost function is extremely complicated and time-consuming to evaluate. As a result, some care must be taken to minimize the number of times a cost is calculated. One step toward reducing the number of function evaluations is to insure that identical chromosomes are not evaluated more than once. There are several approaches to avoiding twins. First, the initial population can be created with no two chromosomes alike. Generally, this is only a problem for the binary genetic algorithm, since the continuous parameter genetic algorithm has very low odds of having identical random parameters appear in two different chromosomes. However, checking the random population for repetitions is time-consuming in itself. Fortunately, there are more efficient approaches. One approach is to begin each chromosome with a different pattern. To illustrate, consider an initial population of eight chromosomes:

$$
\begin{array}{l}
000101010 \\
001101010 \\
010010101 \\
011001100 \\
100110011 \\
101000111 \\
110111000 \\
111011011
\end{array}
$$

Observe that the first 3 bits are uniquely prescribed, so each chromosome is guaranteed to be different from all others in the population. Perhaps this approach is not the best, since it only guarantees that the first gene or so is different. A variation is to prescribe some other combination of bits. For example, if the chromosomes are three genes made up of 3 bits each, the first (or some other) bit in each gene could be set, as in

$$
\begin{array}{l}
111111111 \\
100100000 \\
111011100 \\
100000011 \\
000100100 \\
000100000 \\
010010101 \\
011001011
\end{array}
$$

Either approach guarantees an initial population of unique chromosomes but doesn't insure that identical chromosomes are not evaluated in later generations.

Another way to reduce the number of cost function evaluations is to only evaluate the costs of offspring of mutated chromosomes. Nonmutated parents already have an associated cost, so they don't have to be evaluated again. The simplicity of the approach suggests that it should be included in every genetic algorithm whether the cost function is time-consuming to calculate or not. Also, the cost of a chromosome with multiple mutations only needs one evaluation.

A third approach searches the newly formed chromosomes to see if they match other members of the population. Searching the population for identical twins is only worth the effort if the cost function evaluation takes longer than the population search. A new generation consists of nonmutated parents, mutated parents, offspring, and mutated offspring. If there is an identical chromosome that already has an associated cost, then this cost is assigned to the new chromosome. Otherwise, the cost must be calculated from the cost function.

A fourth approach keeps track of every chromosome and cost calculated over all the generations. Any newborn chromosome is compared with other chromosomes on the master list. If it has a twin in the big list, then the cost of that twin is assigned to the chromosome, thus saving the computation of the cost. This approach is the most drastic, being reserved for the most difficult to evaluate cost functions, because the big list can potentially take up considerable memory and the search becomes time-consuming.

There are some other interesting tricks to play with complicated cost functions. One technique is to find a way to simplify the cost function for the bulk of the runs. This approach is akin to doing an initial calculation on a coarse grid, then fine-tuning it on a much finer grid. Here we merely define a simpler or lower order cost function in the initial generations and increase the cost function accuracy as the genetic algorithm progresses through the generations. The lower order model takes less time to calculate and can get the algorithm into the ballpark of the solution. This approach proved very successful in Haupt (1995). In this case, collocation was done at only one-fifth of the grid points. Thus a matrix on the order of 100×100 was used in the early generations, while a matrix on the order of 500×500 was used in the final generations. Another approach is to use the genetic algorithm to find the valley of the global minimum, then enlist the help of a fast local optimizer to find the bottom of the valley. This method only works if the local optimizer can be applied to the cost function.

Some genetic algorithm users advocate populations with all unique members (Michalewicz, 1992; Davis, 1996). New members of the population (offspring or mutated chromosomes) are checked against the chromosomes in the current population. If the new chromosome is a duplicate, then it is discarded. This approach requires searching the population for each new member. An alternative is to not allow identical chromosomes to mate, therefore saving some computer time while helping maintain diversity in the population.

So we have seen that there are a variety of methods to save computer time by minimizing the number of evaluations of the cost function. Just be careful that the method does not require more time for searching than it saves in cost function evaluations.

5.2 GRAY CODES

Ordinary binary number representation of the parameter values may slow convergence of a genetic algorithm. Consider the following example in which the crossover point for two parents falls in the middle of a gene that lies in the middle of the parent chromosomes.

$$parent_1 = [\ldots \underbrace{100 \overset{crossover}{\downarrow} 00000}_{gene} \ldots] = [\ldots 128 \ldots]$$

$$parent_2 = [\ldots \underbrace{011 \overset{crossover}{\downarrow} 11111}_{gene} \ldots] = [\ldots 127 \ldots]$$

The decoded gene appears at the far right-hand side of the equality. Crossover splits the genes as indicated. The offspring and associated values resulting from a crossover point between bits three and four are given by

$$offspring_1 = [\ldots 10011111 \ldots] = [\ldots 159 \ldots]$$

$$offspring_2 = [\ldots 01100000 \ldots] = [\ldots 96 \ldots]$$

By definition, the parents are some of the better chromosomes. This mating resulted in offspring that have diverging cost values from the parents. The parameters of the genes that are split by the crossover operation in the above examples have decimal representations of 128 and 127. Note that the parameter representations and, most likely, the costs are very close, but the binary representations are exactly opposite. Consequently, the offspring that result from the parents are quite different. In this case, the parameter values change to 159 and 96. The parameter values should be converging, but they are not. This example is extreme, but not unlikely. Increasing the number of bits in the parameter representation magnifies the problem (Haupt, 1996).

One way to avoid this problem is to encode the parameter values using a Gray code. The Gray code redefines the binary numbers so that consecutive numbers have a Hamming distance (Taub and Shilling, 1986) of one. Hamming distance is defined as the number of bits by which two chromosomes differ. Reconsider the previous example using a Gray code in place of binary numbers. If we define the binary coding [01000000] to mean the number 127 and [11000000] to denote 128 then the problem of crossover becomes

$$parent_1 = [\dots \underbrace{110 \overset{crossover}{\downarrow} 00000}_{gene} \dots] = [\dots 128 \dots]$$

$$parent_2 = [\dots \underbrace{010 \overset{crossover}{\downarrow} 00000}_{gene} \dots] = [\dots 127 \dots]$$

One gene in the chromosome is split by the crossover operation. The offspring resulting from a crossover point between bits three and four are given by

$$offspring_1 = [\dots 11000000 \dots] = [\dots 128 \dots]$$

$$offspring_2 = [\dots 01000000 \dots] = [\dots 127 \dots]$$

By definition, the parents are good solutions. Thus, we would expect the offspring to be good solutions too—which occurs when a Gray code is used.

A Gray code is easy to implement. Converting a binary code to a Gray code is done by the process diagrammed in Figure 5.1. The converse of

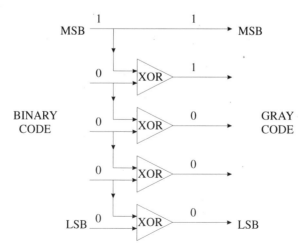

Figure 5.1 Diagram of converting a binary code to a Gray code.

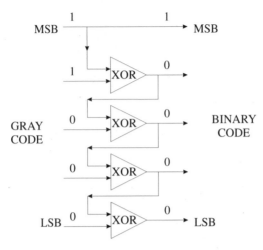

Figure 5.2 Diagram of converting a Gray code to a binary code.

converting a Gray code back to a binary code is accomplished using the inverse process shown in Figure 5.2. Table 5.1 shows the binary and Gray codes for the integers 0 through 7. Note that the Hamming distance is one for every adjacent integer in the Gray code. On the other hand, binary numbers have a Hamming distance of three between integers 3 and 4.

The above argument points out that Gray codes speed convergence time by keeping the algorithm's attention on converging toward a solution. The cost surface exploration is primarily relegated to the mutation function. Our experience indicates that a Gray code slows the genetic algorithm in

TABLE 5.1 Binary and Gray Code Representations of Integers 1 to 7

Integer	Binary Code	Gray Code
0	000	000
1	001	001
2	010	011
3	011	010
4	100	110
5	101	111
6	110	101
7	111	100

time by about 10 to 20% and doesn't usually help much. Others have found Gray codes very beneficial (Caryana and Schaffer, 1988; Hinterding et al., 1989).

5.3 GENE SIZE

In the binary genetic algorithm the size of the gene or the number of bits used to represent a parameter is important to the accuracy of the solution and the length of time needed for the genetic algorithm to converge. Computer users tend to over represent the true accuracy of parameters, because the computer calculates too many digits of accuracy. For every decimal point of accuracy (in base 10) needed in the solution, the parameter needs 3.3 bits in the gene. Thus, a solution accurate to eight positions requires $8 \times 3.3 = 26.4$ or 27 bits in the gene. If there are 10 parameters, then a chromosome has 270 bits. A long gene slows converge of the genetic algorithm. Lesson: Only represent the parameters to the minimum accuracy necessary.

In order to precisely estimate the gene size, the acceptable parameter accuracy in the form of quantization error must first be determined. Tolerances of parts, design specifications, manufacturing limitations, and sufficient numerical accuracy contribute to determining the minimum quantization level. An advantage of the binary genetic algorithm is that it forces the user to determine the accuracy of all parameters in the optimization process. It is foolish to optimize a design only to find that the tolerance of a part is too tight. This is like the government overspecifying tolerances on simple items and driving up the cost. So be careful to avoid the genetic algorithm equivalent of the $500 hammer (or was it a $500 toilet seat?).

5.4 POPULATION

Deciding upon the sizes of the initial and generational populations is very difficult. There are some trade-offs between population sizes and the number of generations needed to converge. Let's look at a worst-case scenario of the number of cost function evaluations needed to find a solution:

$$\#evaluations \leq N_{ipop} + M_{iter} \left[N_{bad} + \mu \left(N_{good} - 1 \right) \right] \qquad (5.1)$$

where M_{iter} is the number of generations. This formula assumes that mutations only occur for the N_{good} chromosomes (excluding the best chromosome) and that any chromosome can only receive a single mutation. By golly, this is a minimization problem! What combination of genetic algorithm parameters result in the minimum number of function evaluations? Grefenstette (1986) used a genetic algorithm to optimize six genetic algorithm parameters. These results are discussed later in this chapter. Unfortunately, the optimum parameters for one problem are not the optimum parameters for another problem.

A small population size causes the genetic algorithm to quickly converge on a local minimum, because it insufficiently samples the parameter space. On the other hand, a large population size takes too long to find and assemble the building blocks to the optimum solution. Goldberg (1986) found that relatively large populations are good for parallel implementations of the genetic algorithm, while relatively small populations are good for serial implementations. Syswerda (1991) offers some advice in choosing population size:

> General wisdom dictates that a larger population will work more slowly but will eventually achieve better solutions than a smaller population. Experience indicates, however, that this rule of thumb is not always true, and that the most effective population size is dependent on the problem being solved, the representation used, and the operators manipulating the representation.

This quote often feeds skeptics with criticism of genetic algorithms.

How should the initial population sample the cost surface? This depends on the type of crossover used and the mutation rate. Algorithms geared toward cost surface exploitation (gradient-based algorithms being the extreme case) need a higher sampling rate than algorithms geared toward cost surface exploration (random search being the extreme case). In order to better picture the sampling process, let's first look at a small population size (16) for a two-parameter cost function. One option is to start with a population that uniformly samples the cost surface (Figure 5.3). Sometimes a random population results in gross oversampling of some regions and sparse sampling in others (Figure 5.3). Uniform random number generators produce uniform samples when the population size is large. An alternative is to insure that each parameter is uniformly sampled while the other parameter(s) is (are) randomly sampled. Figure 5.3 shows the partially random sampling case where each parameter has eight uniform samples. A second alternative is to randomly generate half the chromosomes, then the second half is the complement of the first half (Figure 5.3). This approach insures diversity by requiring every bit to assume both a one and a zero within the population. An example of applying this techinique is:

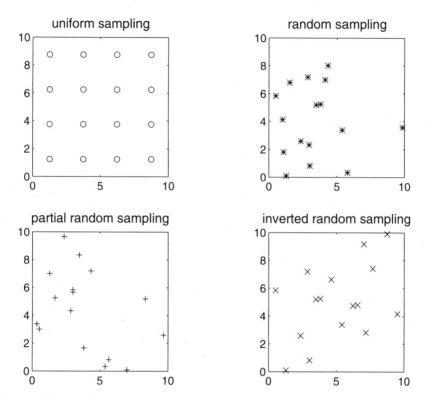

Figure 5.3   Plots of possible sampling methods with 16 samples.

Initial Population of Chromosomes

$$\text{random}\begin{cases} 0 & 1 & 1 & 1 & 1 & 1 & 0 & 0 & 0 & 0 & 1 & 1 & 0 & 0 \\ 0 & 1 & 0 & 1 & 0 & 0 & 1 & 0 & 1 & 0 & 0 & 0 & 0 & 0 \\ 0 & 0 & 1 & 0 & 1 & 0 & 0 & 1 & 0 & 0 & 0 & 1 & 0 & 0 \\ 0 & 0 & 1 & 0 & 1 & 0 & 1 & 1 & 0 & 0 & 0 & 1 & 1 & 0 \end{cases}$$

$$\text{complement}\begin{cases} 1 & 0 & 0 & 0 & 0 & 0 & 1 & 1 & 1 & 1 & 0 & 0 & 1 & 1 \\ 1 & 0 & 1 & 0 & 1 & 1 & 0 & 1 & 0 & 1 & 1 & 1 & 1 & 1 \\ 1 & 1 & 0 & 1 & 0 & 1 & 1 & 0 & 1 & 1 & 1 & 0 & 1 & 1 \\ 1 & 1 & 0 & 1 & 0 & 1 & 0 & 0 & 1 & 1 & 1 & 0 & 0 & 1 \end{cases}$$

Helping the algorithm adequately sample the cost surface in the early stages may reduce convergence time. The initial population size (N_{ipop}) is greater than or equal to the population size from generation to generation (N_{pop}). The rationale for this is that once the cost surface is adequately sampled (explored), the algorithm works with a subset of the best samples

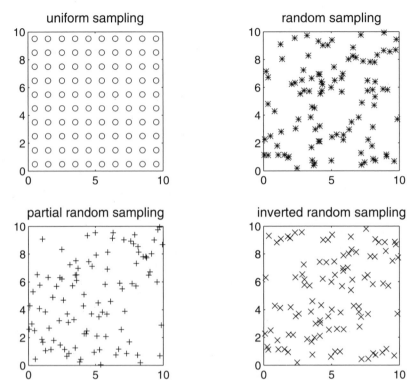

Figure 5.4   Plots of possible sampling methods with 100 samples.

(exploitation). Crossover and mutations increases the algorithm's sampling range.

When the population is sufficiently large, any of the above techniques produces an adequate sampling of the cost surface. Figure 5.4 shows the four sampling methods when there are 100 total samples of the cost surface. All of these samples cover the cost surface well.

We devised a simple check to determine the best population size. The genetic algorithm optimized several functions, and the results are averaged over 100 independent runs. We kept $N_{ipop} = N_{pop}$ and $M_{iter} \times N_{pop} =$ constant (see Table 5.2). The cost functions are the highly undulating

TABLE 5.2   Mutation Rate and Population Size Changed to Keep the Number of Function Evaluations Constant

Run	1	2	3	4	5	6
M_{iter}	160	80	40	20	10	5
N_{pop}	4	8	16	32	64	128

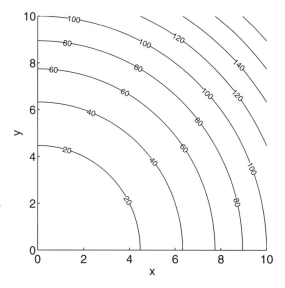

Figure 5.5 Plot of the cost surface of $f(x, y) = x^2 + y^2$.

surface expressed by

$$f(x, y) = x \sin(4x) + 1.1y \sin(2y) \tag{5.2}$$

from Chapters 1 and 3 (see Figures 1.3 and 1.4), and the parabolic function (Figure 5.5)

$$f(x, y) = x^2 + y^2 \tag{5.3}$$

Both are constrained to the region $0 \le x \le 10$ and $0 \le y \le 10$. These functions test the genetic algorithm for both highly oscillatory and smooth cost surfaces. First, consider the continuous parameter genetic algorithm. A graph of the minimum cost vs. run number function of equation (5.2) is shown in Figure 5.6. Contrast these results with those for the equation (5.3) cost function (Figure 5.7). In both cases, a small population size ($N_{pop} = 8$) and many generations ($M_{iter} = 80$) produce the best results. These same trials were repeated with a binary genetic algorithm (Figures 5.8 and 5.9). The binary genetic algorithm worked best for $N_{pop} = 4$ and $M_{iter} = 160$—again, extremely small population sizes. These results were similar for a range of mutation rates.

Changing the population size from generation to generation is another variation that we haven't explored, yet is very common in nature. Another

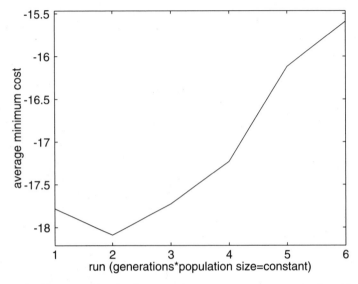

Figure 5.6 Convergence plot for a continuous parameter genetic algorithm with the cost function in equation (5.2) as a function of the population size.

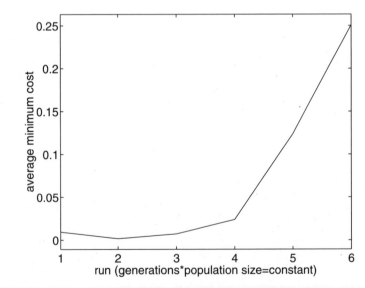

Figure 5.7 Convergence plot for a continuous parameter genetic algorithm with the cost function in equation (5.3) as a function of the population size.

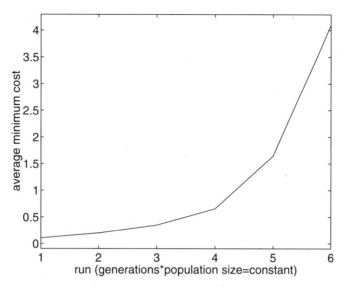

Figure 5.8 Convergence plot for a binary genetic algorithm with the cost function in equation (5.2) as a function of the population size.

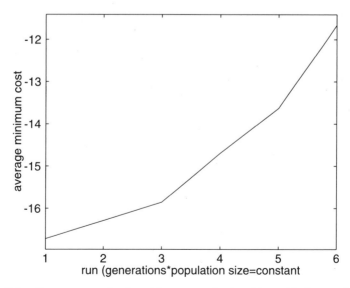

Figure 5.9 Convergence plot for a binary genetic algorithm with the cost function in equation (5.3) as a function of the population size.

intriguing biological model is allowing a chromosome to age (Michalewicz, 1992). A chromosome's age equals the number of generations that it survives. More fit chromosomes stay alive longer than less fit chromosomes. Such a strategy allows the population size to vary from generation to generation, depending on the number of chromosome deaths. Michalewicz lists three possibilities of adding a lifetime to the chromosomes: proportional, linear, and bilinear. Their definitions are presented in Table 5.3. The variables are defined as

- $life_{min}$ = minimum lifetime
- $life_{max}$ = maximum lifetime
- $\eta = \frac{1}{2}(life_{max} - life_{min})$
- $cost_{min}$ = minimum cost of a chromosome in the population
- $cost_{max}$ = maximum cost of a chromosome in the population
- $E\{cost\}$ = expected value of the cost vector

He found that the varying population size outperformed the constant population size genetic algorithm for the cost functions shown in Table 5.4. Results consisted of averaging the minimum cost and number of function evaluations over 20 independent runs. The linear strategy has the best average minimum cost but the highest number of function evaluations. At the other extreme, the bilinear strategy has the worst average minimum cost but the smallest number of function evaluations. The proportional strategy came in the middle with a medium average minimum cost and medium number of function evaluations.

Is it always best to keep half of the chromosomes each generation? To assess this issue, we varied N_{good} from 2 to 40 with $N_{ipop} = N_{pop} = 48$ and $\mu = 0.2$. Again, the continuous parameter genetic algorithm outperformed the binary one when the minimum cost per generation was averaged over 100 independent runs. Results for the cost functions in equations (5.2) and (5.3) are shown in Figures 5.10 and 5.11 for the continuous parameter genetic algorithm, and in Figures 5.12 and 5.13 for the binary genetic algorithm. The continuous parameter genetic algorithm performed best on both cost functions when half or less of the chromosomes were kept from generation to generation. Contrast this performance with the binary genetic algorithm where the best N_{good} depends on the cost function. The binary genetic algorithm works best with a small value for N_{good} for a smooth cost function and a large value for N_{good} for the oscillatory cost function. At least for these versions of a genetic algorithm and cost functions, keeping about half of the population from generation to generation seems to be a good idea.

TABLE 5.3 Comparison of Three Ways of Adding a Life Span to a Chromosome

Allocation	$Lifetime_i$	Advantage	Disadvantages
Proportional	$\min\left\{life_{max},\ life_{max} - \eta\dfrac{cost_i}{E\{cost\}}\right\}$	$lifetime_i \propto \dfrac{1}{cost}$	• $lifetime_i$ relative to average not best cost • $cost_i \geq 0$ for all i
Linear	$life_{max} - 2\eta\dfrac{cost_i - cost_{min}}{cost_{max} - cost_{min}}$	Compares to best cost	chromosomes with long lives
Bilinear	$\begin{cases} life_{min} + \eta\dfrac{cost_{max} - cost_i}{cost_{max} - E\{cost\}} & \text{if } cost_i \geq E\{cost\} \\[2ex] \dfrac{1}{2}\left(life_{min} + life_{max}\right) + \eta\dfrac{E\{cost\} - cost_i}{E\{cost\} - cost_{min}} & \text{if } cost_i < E\{cost\} \end{cases}$	Good compromise	—

TABLE 5.4   Functions Used by Michalewicz to Test the Performance of a Genetic Algorithm that has Chromosomes with a Life Span

Number	Function	Limits
1	$-x\sin(10\pi x) + 1$	$-2.0 \leq x \leq 1.0$
2	$integer(8x)/8$	$0.0 \leq x \leq 1.0$
3	$x \cdot sgn(x)$	$-1.0 \leq x \leq 2.0$
4	$0.5 + \dfrac{\sin^2 \sqrt{x^2 + y^2} - 0.5}{\left(1 + 0.001\left(x^2 + y^2\right)\right)^2}$	$-100 \leq x \leq 100$

Will seeding the population with some possible good guesses speed convergence? Sometimes. This question (and answer) is natural since some traditional methods require a good first guess to get a good solution. If the guess is very good, the algorithm quickly finds the solution. If the guess is not so good, then the algorithm chases after a local minimum and takes time to find its way out. Most iterative schemes have a similar problem with initial first guesses (Haupt, 1987). If you don't know much about the expected best solution (which is usually the case), forget about seeding the population. Another problem with seeding is that the time spent looking for

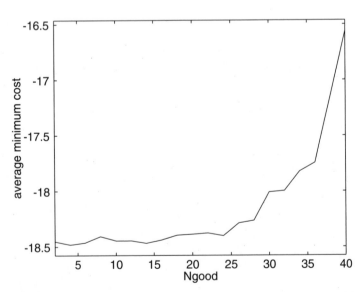

Figure 5.10   Convergence plot for a continuous parameter genetic algorithm with the cost function in equation (5.2) as a function of N_{good}.

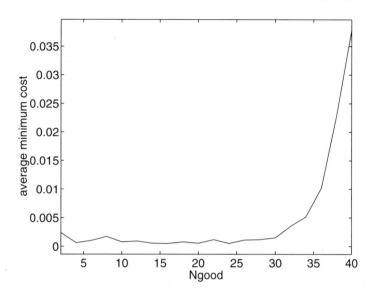

Figure 5.11 Convergence plot for a continuous parameter genetic algorithm with the cost function in equation (5.3) as a function of N_{good}.

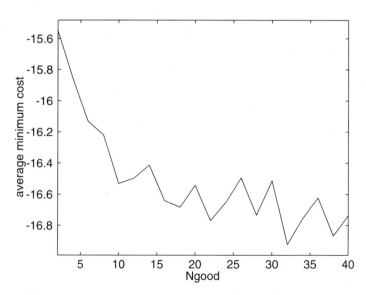

Figure 5.12 Convergence plot for a binary genetic algorithm with the cost function in equation (5.2) as a function of N_{good}.

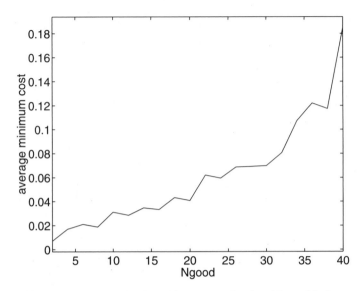

Figure 5.13 Convergence plot for a binary genetic algorithm with the cost function in equation (5.3) as a function of N_{good}.

a good seed can oftentimes be spent running the genetic algorithm. We're not against seeding a genetic algorithm—we sometimes do it ourselves—we're just cautious. Others have not found stunning results with seeding, either (Booker, 1987; Liepens et al., 1990).

5.5 CONVERGENCE

Do genetic algorithms converge? There is no mathematical proof of convergence or any guarantee that they find the global minimum. There is a handwaving proof for the binary genetic algorithm called the schema theorem (Holland, 1975). A schema is a string of characters consisting of the binary digits 1 and 0, and an additional "don't care" character, #. Thus, the schema 11##00 means the center two digits can be either a 1 or a 0 and represents the four strings given by 110000, 110100, 111000, and 111100. The schema theorem says:

> Short schema with better than average costs occur exponentially more frequently in the next generation. Schema with costs worse than average occur less frequently in the next generation.

The idea is that the most fit schema survive to future generations. Following the best schema throughout the life of the genetic algorithm provides an estimate of the convergence rate (not necessarily to the global minimum) to the best chromosome.

Say that a given schema has s_t representations in the population at generation t. The number of representations of this schema in generation $t + 1$ is given by

$$s_{t+1} = s_t P_t (1 + Q_t) R_t \tag{5.4}$$

where

> P_t = probability of the schema being selected to survive to the next generation
>
> Q_t = probability of the schema being selected to mate
>
> R_t = probability of the schema not being destroyed by crossover or mutation.

Similar formulas are presented in many references (e.g., Goldberg, 1989a). Notice that if $P_t (1 + Q_t) < 1$, then the schemata will eventually become extinct. Surviving schema propagate according to

$$s_{t+1} \geq (P_t P_{t-1} \cdots P_1) [(1 + Q_t)(1 + Q_{t-1}) \cdots (1 + Q_1)] (R_t R_{t-1} \cdots R_1) s_1 \tag{5.5}$$

This formula takes into account that the probability of selection of a schema can vary from generation to generation. Beginning generations may have a schema with $P_t (1 + Q_t) > 1$, and for later generations that same schema has $P_t (1 + Q_t) < 1$. Such a schema does not survive in the long run. An example of this behavior might be when a genetic algorithm first latches onto a local minimum, then later finds the global minimum. Ultimate survival requires the schema to have $P_t (1 + Q_t) > 1$ for all t.

In practice when do you stop the algorithm? Well ... We don't have a good answer. This is one of the fuzzy parts of using the genetic algorithm. Some good possibilities:

1. Correct answer. This may sound silly and simple. Make sure you check your best chromosome to see if it is the answer or an acceptable answer to the problem. If it is, then stop.

2. No improvement. If the genetic algorithm continues with the same best chromosome for X number of iterations, then stop. Either the algorithm found a good answer or it is stuck in a local minimum. Be

careful here. Sometimes the best solution will stay in the same spot for quite a while before a fortuitous crossover or mutation generates a better one.

3. Statistics. If the mean or standard deviation of the population's cost reaches a certain level, then stop the algorithm. This means the values are no longer changing.

4. Set number of iterations. If the algorithm doesn't stop for one of the above reasons, then limit the maximum number of iterations. It can go on forever unless stopped.

If your algorithm isn't converging to a good solution, try changing the genetic algorithm variables like population size and mutation rate. Maybe a different crossover or switching from a continuous parameter genetic algorithm to a binary genetic algorithm is the answer. We do not advocate the genetic algorithm as an answer to every problem. Sometimes it performs poorly in comparison with other methods. Don't be afraid to jump ship to a minimum seeking algorithm when your genetic algorithm is sinking.

5.6   ALTERNATIVE CROSSOVERS FOR BINARY GENETIC ALGORITHMS

In Chapter 2 we only introduced the simple single point crossover for binary genetic algorithms. Other alternatives exist. A two-point crossover takes the form

$$
\begin{array}{ll}
parent_1 & 001\ \overbrace{01011000}110 \\
parent_2 & 011\ \underbrace{1110000}1100 \\
offspring_1 & 001\ \underbrace{1110000}0110 \\
offspring_2 & 011\ \overbrace{01011001}100
\end{array}
$$

Two random crossover points are selected for the parents. The parents then swap the bits between the two crossover points. Alternatively, a random selection of one of the three parts of the chromosome determines which bits are swapped. The previous example occurs when 2 is the random number. If 3 is the random number, the two parents swap the last 5 bits in their chromosomes.

Another scheme involves three parents and two crossover points.

$$
\begin{array}{ll}
parent_1 & 0101\mathbf{01010}\mathit{10101} \\
parent_2 & 1111\mathbf{1110}00000 \\
parent_3 & 1100\mathbf{1100}1\mathit{10011} \\
offspring_1 & 0101\mathbf{01010}00000 \\
offspring_2 & 0101\mathbf{01010}\mathit{10011} \\
offspring_3 & 1111\mathbf{1110}0\mathit{10101} \\
offspring_4 & 1111\mathbf{1110}0\mathit{10011} \\
offspring_5 & 1100\mathbf{1100}1\mathit{10101} \\
offspring_6 & 1100\mathbf{1100}100000 \\
\vdots & \vdots \\
offspring_{18} & 1111\mathbf{01010}\mathit{10011}
\end{array}
\qquad (5.6)
$$

A total of 18 offspring can be generated from the three parents. Not all these offspring need to be generated. For instance, it may be desirable to generate only three. Using the previous scheme of swapping randomly chosen parts of the chromosome is usually appropriate.

Uniform crossover looks at each bit in the parents and randomly assigns the bit from one parent to one offspring and the bit from the other parent to the other offspring. First a random mask is generated. This mask is a random vector of ones and zeros and is the same length as the parents. When the bit in the mask is 0, then the corresponding bit in $parent_1$ is passed to $offspring_1$ and the corresponding bit in $parent_2$ is passed to $offspring_2$. When the bit in the mask is 1, then the corresponding bit in $parent_1$ is passed to $offspring_2$ and the corresponding bit in $parent_2$ is passed to $offspring_1$:

$$
\begin{array}{ll}
parent_1 & 00101011000110 \\
parent_2 & 01111100001100 \\
mask & 00110110001110 \\
offspring_1 & 00111101001100 \\
offspring_2 & 01101010000110
\end{array}
$$

Uniform crossover becomes single-point crossover for a mask like

$$mask = [00000011111111]$$

In the same manner, a two-point crossover mask example is

$$mask = [00000011111000]$$

Thus, uniform crossover can be considered a generalization of the other crossover methods. Syswerda (1989) has done extensive experimentation with uniform crossover. "It was shown that in almost all cases, uniform crossover is more effective at combining schemata than either one- or two-point crossover." In addition, he found that two-point crossover is consistently better than one-point crossover.

One can also vary how many of the N_{good} chromosomes become parents. The crossover rate (X_{rate}) determines the number of chromosomes that enter the mating pool.

$$N_{good} = X_{rate}N_{pop} \tag{5.7}$$

High crossover rates introduce many new chromosomes into the population. If X_{rate} is too high, good building blocks don't have the opportunity to accumulate within a single chromosome. A low X_{rate}, on the other hand, doesn't do much exploring of the cost surface.

The above schemes provide alternative implementations of crossover. Of course, there are many other possibilities. Feel free to use your imagination to invent other methods. Just remember, the goal of the crossover operator is to pass on desirable traits to the next generation. A crossover operator that is too fancy may destroy desirable schema and slow convergence.

5.7 MUTATION

When reevaluating implementation of the mutation operator, two issues come to mind: type of mutation and rate of mutation.

How severe should the mutation be? Changing a single bit in a gene can change a parameter value by almost 50%. The expected value of a mutated gene that represents a parameter between 0 and 1 to N bit accuracy is $1/N \sum_{n=1}^{N} 2^{-n}$. Thus a gene with 4 bits can expect a mutation ($\mu = 0.25$) to change it by $\frac{1}{4}(0.5 + 0.25 + 0.125 + 0.0675) = 0.23563$, while a gene with 8 bits can expect two mutations ($\mu = 0.25$), and it changes by $\frac{2}{8} \left(\frac{1}{2} + \frac{1}{2^2} + \frac{1}{2^3} + \frac{1}{2^4} + \frac{1}{2^5} + \frac{1}{2^6} + \frac{1}{2^7} + \frac{1}{2^8} \right) = 0.24902$. In other words, mutation has slightly different effects on genes, depending on the bit representation of the genes.

The genetic algorithm is sensitive to the choice of mutation rate. Figure 5.14 shows a graph of the minimum cost per generation (off-line performance) found by the continuous parameter genetic algorithm after 25 generations averaged over 100 runs versus mutation rate, μ for equation (5.2). Likewise, Figure 5.15 shows results for equation (5.3). At least for these two cases, the best range of μ is between is 0.14 and 0.26 for the highly oscillatory function of equation (5.2), and between 0.06 and 0.34 for

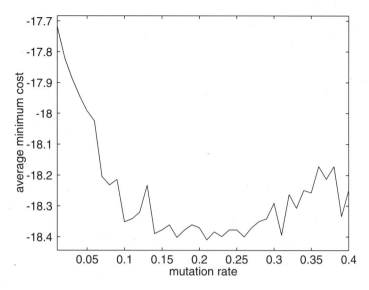

Figure 5.14 Convergence plot for a continuous parameter genetic algorithm with the cost function in equation (5.2) as a function of μ.

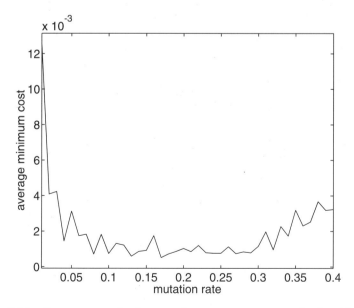

Figure 5.15 Convergence plot for a continuous parameter genetic algorithm with the cost function in equation (5.3) as a function of μ.

the smooth function, equation (5.3). Similar graphs for the binary genetic algorithm appear in Figures 5.16 and 5.17. The binary genetic algorithm works best for $0.1 \leq \mu \leq 0.4$. These results are contrary to the suggested small mutation rates in the literature (DeJong, 1975; Grefenstette, 1986).

Experiments have been done on varying the mutation rate as the generations progress. Fogarty studied five variable mutation rate schemes (Fogarty, 1989):

1. Constant low μ over all generations.
2. First generation has $\mu = 0.5$, and subsequent generations have a low μ.
3. Exponentially decreasing μ over all generations.
4. Constant μ for all bits.
5. Exponentially decreasing μ for bits representing small inverse powers of 2.

Fogarty tested these schemes on a model of an industrial burner in which the air inlet valves were set in order to minimize combustion stackloss in the common flue. The conclusions indicated that the variable mutation rate worked better than the constant mutation rate when the initial population was a matrix of all zeros. When the initial population was a matrix of random bits, then no significant difference between the various mutation rates was noticed. Since these results were performed on one specific problem, no general conclusions about a variable mutation rate are possible.

5.8 PERMUTATION PROBLEMS

Sometimes an optimization involves sorting a list or putting things in the right order. The most famous problem is the traveling salesman problem in which a salesman wants to visit C cities while traveling the least possible distance. Each city is visited exactly once, so the solution consists of a list of the cities in the order visited. This particular application is treated in more detail in the next chapter, but let's take a look at the problems involved and some possible solutions. In particular, the standard crossover and mutation operators are not appropriate since we must insure that each city is represented once and only once.

Let's start with a simple example of six numbers that must be reordered. For simplicity, we will use an integer alphabet here although any reasonable encoding will do. Consider two parent chromosomes of length 6.

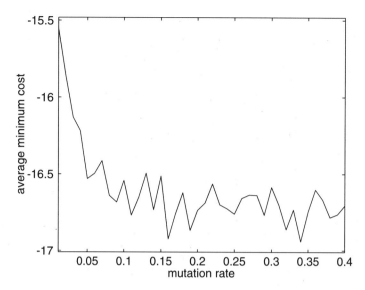

Figure 5.16 Convergence plot for a binary genetic algorithm with the cost function in equation (5.2) as a function of μ.

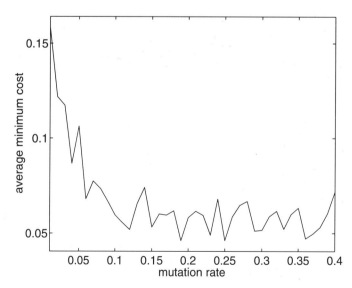

Figure 5.17 Convergence plot for a binary genetic algorithm with the cost function in equation (5.3) as a function of μ.

Parents	
*parent*₁	[3 4 6 2 1 5]
*parent*₂	[4 1 5 3 2 6]

A simple crossover between the second and third elements of the parent chromosome vectors produces the offspring:

Simple Crossover (incorrect)	
*offspring*₁	[3 4	5 3 2 6]
*offspring*₂	[4 1	6 2 1 5]

Obviously this won't work since *offspring*₁ contains two 3s and no 1, while *offspring*₂ has two 1s, and no 3. Goldberg (1989) discusses several possible solutions to this problem, which are briefly summarized here. The first is known as partially matched crossover (PMX) (Goldberg and Lingle, 1985). In this method two crossover points are chosen, and we begin by exchanging the values of the parents between these points. If crossover points are between elements 1 and 2, and 3 and 4, then string *K* from *parent*₂ is switched with string *J* from *parent*₁.

Partially Matched Crossover (step A)	
*offspring*₁ₐ	[3	**15**	2 1 5]
	J
*offspring*₂ₐ	[4	**46**	3 2 6]
	K

All values exchanged between parents are shown in bold type. So far we still have the problem of having doubles of some integers and none of others. The switched strings, *J* and *K*, remain untouched throughout the rest of the procedure. The original doubles in *offspring*₂ₐ are exchanged for the original doubles in *offspring*₁ₐ (the original 4 in *offspring*₂ₐ exchanged with the original 1 in *offspring*₁ₐ, and the original 6 in *offspring*₂ₐ with the original 5 in *offspring*₁ₐ) to obtain the final solution.

Partially Matched Crossover (step B)	
*offspring*₁ᵦ	[3 1 5 2 4 6]
*offspring*₂ᵦ	[1 4 6 3 2 5]

Each offspring contains part of the initial parent in the same position (unemphasized numbers) and includes each integer once and only once.

Order crossover (OX) is somewhat different than PMX. It attempts to maintain the order of integers as if the chromosome vector were wrapped in a circle, so the last element is followed by the first element. Thus, [1 2 3 4] is the same as [2 3 4 1]. It begins, like PMX, by choosing two crossover points and exchanging the integers between them. This time, however, holes are left (denoted below by X's) in the spaces where integers are repeated. If the crossover points are after the second and fourth integers, the first stage leaves offspring that look like:

Ordered Crossover (First Stage)	
$offspring_{1L}$	$[X\ 4\ \|\ \underbrace{5\ 3}_{J}\ \|\ 1\ X]$
$offspring_{2L}$	$[4\ 1\ \|\ \underbrace{6\ 2}_{K}\ \|\ X\ X]$

At this point the holes are pushed to the beginning of the offspring. All integers that were in those positions are pushed off the left of the chromosome and wrap around to the end of the offspring. At the same time, strings J and K that were exchanged maintain their positions.

Ordered Crossover (Second Stage)	
$offspring_{1M}$	$[X\ X\ 5\ 3\ 1\ 4]$
$offspring_{2M}$	$[X\ X\ 6\ 2\ 4\ 1]$

For the final stage, the X's are replaced with strings J and K.

Ordered Crossover (Final Stage)	
$offspring_{1N}$	$[\underbrace{6\ 2}_{K}\ \underbrace{5\ 3}_{J}\ 1\ 4]$
$offspring_{2N}$	$[\underbrace{5\ 3}_{J}\ \underbrace{6\ 2}_{K}\ 4\ 1]$

OX has the advantage that the relative ordering is preserved, although the absolute position within the string is not.

The final method discussed by Goldberg (1989) is the cycle crossover (CX). In this method, the information exchange begins at the left of the string and the first two digits are exchanged. For our example, this gives:

Cycle Crossover (First Step)	
$offspring_{1W}$	[4 4 6 2 1 5]
$offspring_{2W}$	[3 1 5 3 2 6]

Now that the first offspring has two 4s, we progress to the second 4 and exchange with the other offspring to get

Cycle Crossover (Second Step)	
$offspring_{1X}$	[4 1 6 2 1 5]
$offspring_{2X}$	[3 4 5 3 2 6]

Now there are two 1s in the first offspring, so we exchange position 5 with the second offspring.

Cycle Crossover (Third Step)	
$offspring_{1Y}$	[4 1 6 2 2 5]
$offspring_{2Y}$	[3 4 5 3 1 6]

The next position to exchange is position 4 where there is a repeated 2.

Cycle Crossover (Fourth Step)	
$offspring_{1Z}$	[4 1 6 3 2 5]
$offspring_{2Z}$	[3 4 5 2 1 6]

At this point, we have exchanged the 2 for the 3 and are back to our starting point, so the crossover is complete. We see that each offspring has exactly one of each digit, and it is in the position of one of the parents. Comparing the three different methods, we see that each has produced a different set of offspring, offspring B, N, and Z. These methods are compared by Oliver et al. (1987).

What about the mutation operator? If we randomly change one number in a string, we are left with one integer duplicated and another missing. The simplest solution is to randomly choose a chromosome to mutate, then randomly choose two positions within that chromosome to exchange. As an example, if the second and fifth positions are randomly chosen for a mutated chromosome, it transforms as follows:

$$chromosome = [6\,\mathbf{1}\,5\,3\,\mathbf{2}\,4]$$
$$\Downarrow$$
$$chromosome = [6\,\mathbf{2}\,5\,3\,\mathbf{1}\,4]$$

So we see that even permutation problems are not an insurmountable problem for a genetic algorithm. In the following chapter we demonstrate its application.

5.9 SELECTING GENETIC ALGORITHM PARAMETERS

Selecting genetic algorithm parameters like mutation rate, μ, and population size, N_{pop}, is very difficult due to the many possible variations in the algorithm. A genetic algorithm relies upon random number generators for generating the population, mating, and mutation. A different random number seed produces different results. In addition, there are various types of crossovers and mutations, as well as other possibilities, like chromosome aging and Gray codes. Comparing all the different options and averaging the results to reduce random variations for a wide range of cost functions is a daunting task. Plus, the results may be highly dependent on the cost function analyzed.

The first intensive study of genetic algorithm parameters was done by De Jong (1975) in his dissertation. His work translated Holland's theories into practical function optimization. Goldberg (1989) nicely summarizes this work. De Jong used two performance measures for judging the genetic algorithms. First, he defined on-line performance as an average of all costs up to the present generation. It penalizes the algorithm for too many poor costs and rewards the algorithm for quickly finding where the lowest costs lie. Second, he defined off-line performance as the best cost found up to the present generation. This metric doesn't penalize the algorithm for exploring high cost areas of the cost surface. The binary genetic algorithms were of six types with an increasing degree of complexity:

1. A simple genetic algorithm was used, composed of roulette wheel selection, simple crossover with random mating, and single bit mutation.
2. Elitism was added.
3. Reproduction was a function of the expected number of offspring of a chromosome.

4. Numbers 2 and 3 were combined.
5. A crowding factor was added. A small random group from the population was selected. The chromosome in that group that most resembles the offspring was eliminated and replaced by the offspring.
6. Crossovers were made more complex.

Each of these algorithms was tested on five cost functions (shown in Table 5.5) while varying μ, N_{pop}, X_{rate}, and G, where G is the generation gap and has the bounds $0 < G \leq 1$. A generation gap algorithm picks GN_{pop} members for mating. The GN_{pop} offspring produced replace GN_{pop} chromosomes randomly selected from the population.

De Jong concluded:

1. Small population size improved initial performance while large population size improved long-term performance. $50 \leq N_{pop} \leq 100$
2. High mutation rate was good for off-line performance, while low mutation rate was good for on-line performance. $\mu = 0.001$
3. Crossover rates should be about $X_{rate} = 0.60$.
4. Type of crossover (single point versus multipoint) made little difference.

A decade later brought significant improvements to computers that led to the next important study done by Grefenstette (1986). He used a metagenetic algorithm to optimize the on-line and off-line performance of genetic algorithms based on varying six parameters: N_{pop}, X_{rate}, μ, G, scaling win-

TABLE 5.5 Functions Used by Michalewicz to Test the Performance of a Genetic Algorithm that Has Chromosomes with a Life Span

Number	Function	Limits
1	$\sum_{n=1}^{3} x_n^2$	$-5.12 \leq x_n \leq 5.12$
2	$100\left(x_1^2 - x_2\right)^2 + (1 - x_1)^2$	$-2.048 \leq x_n \leq 2.048$
3	$\sum_{n=1}^{5} \mathbf{integer}(x_n)$	$-5.12 \leq x_n \leq 5.12$
4	$\sum_{n=1}^{30} n x_n^4 + \mathbf{Gauss}(0, 1)$	$-1.28 \leq x_n \leq 1.28$
5	$0.002 + \sum_{m=1}^{25} \dfrac{1}{m + (x_1 - a_{1m})^6 + (x_2 - a_{2m})^6}$	$-65.536 \leq x_n \leq 65.536$

dow, and whether elitism is used or not. A scaling window just determines how a cost is scaled relative to other costs in the population. The metagenetic algorithm used $N_{pop} = 50$, $X_{rate} = 0.6$, $\mu = 0.001$, $G = 1$, no scaling, and elitism. These values were chosen based on past experience. A cost function evaluation for the metagenetic algorithm consisted of a genetic algorithm run until 5000 cost function evaluations were performed on one of the De Jong test functions and the result normalized relative to that of a random search algorithm. Each genetic algorithm in the population evaluated each of the De Jong test functions.

The second step in this experiment took the 20 best genetic algorithms found by the metagenetic algorithm and let them tackle each of the five test functions for five independent runs. The best genetic algorithm for on-line performance had $N_{pop} = 30$, $X_{rate} = 0.95$, $\mu = 0.01$, $G = 1$, scaling of the cost function, and elitism. Grefenstette found that the off-line performance was worse than that of a random search, indicating that the genetic algorithms tended to prematurely converge on a local minimum. The best off-line genetic algorithm had $N_{pop} = 80$, $X_{rate} = 0.45$, $\mu = 0.01$, $G = 0.9$, scaling of the cost function, and no elitism. He concludes, "The experimental data also suggests that, while it is possible to optimize GA control parameters, very good performance can be obtained with a range of GA control parameter settings."

A couple of years after the Grefenstette study, a group reported results on optimum parameter settings for a binary genetic algorithm using a Gray code (Schaffer et al., 1989). Their approach added five more cost functions to the De Jong test function suite. They had discrete sets of parameter values ($N_{pop} = 10, 20, 30, 50, 100, 200$; $\mu = 0.001, 0.002, 0.005, 0.01, 0.02,$ 0.05, 0.10; $X_{rate} = 0.05$ to 0.95 in increments of 0.10; and 1 or 2 crossover points) that had a total of 8400 possible combinations. Each of the 8400 combinations was run with each of the test functions. Each combination was averaged over 10 independent runs. The genetic algorithm terminated after 10,000 function evaluations. These authors found the best on-line performance resulted for the following parameter settings: $N_{pop} = 20$ to 30, $X_{rate} = 0.75$ to 0.95, $\mu = 0.005$ to 0.01, and two-point crossover.

Parameter settings are sensitive to the cost functions, options in the genetic algorithms, bounds on the parameters, and performance indicators. Consequently, different studies result in different conclusions about the optimum parameter values. Davis (1989) recognized this problem and outlined a method of adapting the parameter settings during a run of a genetic algorithm (Davis, 1991a). He does this by including operator performance in the cost. Operator performance is the cost reduction caused by the operator divided by the number of children created by the operator.

5.10   CONTINUOUS VS. BINARY GENETIC ALGORITHM

Perhaps a burning question on your mind is whether the binary or continuous parameter genetic algorithm is better. Our experience puts our vote with the continuous parameter genetic algorithm. Figures 5.6 to 5.15 in this chapter indicate that the continuous parameter genetic algorithm outperforms the binary genetic algorithm for both the smooth and highly oscillatory cost functions. This is not always the case, as shown in the antenna array example in Chapter 4. Converting parameter values to binary numbers and worrying about the number of bits needed to represent a parameter are unnecessary. Continuous parameter genetic algorithms also are more compatible with other optimization algorithms, thus making them easier to combine or hybridize.

We are not the only ones to reach this conclusion. After extensive comparisons between binary and continuous parameter genetic algorithms, Michalewicz (1992) says, "The conducted experiments indicate that the floating point representation is faster, more consistent from run to run, and provides a higher precision (especially with large domains where binary coding would require prohibitively long representations)." The inventors of evolutionary computations in Europe have long recognized the value of using continuous parameters in the algorithm.

In any event, both versions of the genetic algorithm are powerful. It is not difficult to find advocates for each.

BIBLIOGRAPHY

Booker, L., 1987, "Improving search in genetic algorithms," in L. Davis, (Ed.) *Genetic Algorithms and Simulated Annealing*, London: Pitman, pp. 61–73.

Caruana, R. A., and J. D. Schaffer, 1988, "Representation and hidden bias: Gray vs. binary coding for genetic algorithms," *Proc. of the 5th International Conference on Machine Learning*, Los Altos, CA: Morgan Kaufmann, June 12–14, pp. 153–161.

Davis, L., 1989, "Adapting operator probabilities in genetic algorithms," in J. D. Schaffer (Ed.), *Proc. of the Third International Conference on Genetic Algorithms*, Los Altos, CA: Morgan Kaufmann, pp. 61–67.

Davis, L., 1991a, "Parameterizing a genetic algorithm," in L. Davis (Ed.), *Handbook of Genetic Algorithms*, New York: Van Nostrand Reinhold.

Davis, L., 1991b, "Performance enhancements," in L. Davis, (Ed.), *Handbook of Genetic Algorithms*, New York: Van Nostrand Reinhold.

De Jong, K. A., 1975, "Analysis of the behavior of a class of genetic adaptive systems," Ph.D. Dissertation, The University of Michigan, Ann Arbor, MI.

Fogarty, T. C., 1989, "Varying the probability of mutation in the genetic algorithm," in J. D. Schaffer (Ed.), *Proc. of the Third International Conference on Genetic Algorithms*, Los Altos, CA: Morgan Kaufmann, pp. 104–109.

Goldberg, D. E., 1989a, *Genetic Algorithms in Search, Optimization, and Machine Learning*, Reading, MA: Addison-Wesley.

Goldberg, D. E., 1989b, "Sizing populations for serial and parallel genetic algorithms," in J. D. Schaffer (Ed.), *Proc. of the Third International Conference on Genetic Algorithms*, Los Altos, CA: Morgan Kaufmann, pp. 70–79.

Goldberg, D. E. and Lingle, R., 1985, "Alleles, loci, and the traveling salesman problem," *Proc. of an International Conference on Genetic Algorithms and their Applications*, pp. 154–159.

Grefenstette, J. J., 1986 Jan./Feb., "Optimization of control parameters for genetic algorithms," *IEEE Trans. Systems, Man, and Cybernetics*. **SMC 16**, p. 128.

Haupt, R. L., 1987, "Synthesis of resistive tapers to control scattering patterns of strips," The University of Michigan, Ann Arbor, MI.

Haupt, R. L., 1995 March, "Optimization of aperiodic conducting grids," *11th Annual Review of Progress in Applied Computational Electromagnetics Conference*, Monterey, CA.

Haupt, R. L., 1996 March, "Speeding convergence of genetic algorithms for optimizing antenna arrays," *12th Annual Review of Progress in Applied Computational Electromagnetics Conference*, Monterey, CA.

Hinterding, R., H. Gielewski, and T. C. Peachey, 1989, "The nature of mutation in genetic algorithms," in L. J. Eshelman (Ed.), *Proc. of the Sixth International Conference on Genetic Algorithms*, Los Altos, CA: Morgan Kaufmann, pp. 70–79.

Holland, J. H., 1975, *Adaptation in Natural and Artificial Systems*, Ann Arbor: The University of Michigan Press.

Liepins, G. E., et al., 1990, "Genetic algorithm applications to set covering and traveling salesman problems," in Brown (Ed.) *OR/AI: The Integration of Problem Solving Strategies*.

Michalewicz, Z., 1992, *Genetic Algorithms + Data Structures = Evolution Programs*, New York: Springer-Verlag.

Oliver, I. M., D. J. Smith, and J. R. C. Holland, 1987, "A study of permutation crossover operators on the traveling salesman problem," *Genetic Algorithms and their Applications: Proc. of the Second International Conference on Genetic Algorithms*, pp. 224–230.

Schaffer, J. D., et al., 1989 June, "A study of control parameters affecting online performance of genetic algorithms for function optimization," in J. D. Schaffer (Ed.), *Proc. of the Third International Conference on Genetic Algorithms*, Los Altos, CA: Morgan Kaufmann, pp. 51–60.

Syswerda, G., 1989 June, "Uniform crossover in genetic algorithms," in J. D. Schaffer (Ed.), *Proc. of the Third International Conference on Genetic Algorithms*, Los Altos, CA: Morgan Kaufmann, pp. 2–9.

Syswerda, G., 1991, "Schedule optimization using genetic algorithms," in L. Davis (Ed.), *Handbook of Genetic Algorithms*, New York: Van Nostrand Reinhold, p. 347.

Taub, H., and D. L. Schilling, 1986, *Principles of Communication Systems*, McGraw-Hill: New York.

CHAPTER 6

ADVANCED APPLICATIONS

Now that we have seen some basic applications of both binary and continuous genetic algorithms and discussed some of the fine points of their implementation, it will be fun to look at what can be accomplished with a genetic algorithm and a bit of imagination. The examples in this chapter make use of some of the advanced topics discussed in Chapter 5 and add variety to the examples presented in Chapter 4. They cover a wide range of areas including technical as well as nontechnical examples. The first example is the infamous traveling salesman problem where the genetic algorithm must order the cities visited by the salesman. The second example revisits the locating-an-emergency-response unit from Chapter 4, but this time uses a Gray code. Next comes a search for an alphabet that decodes a secret message. Our next examples come from engineering design and include robot trajectory planning and introductory stealth design. Finally, we end with two examples from science: one builds an inverse model of a phase space curve and the other finds solutions to a nonlinear fifth-order differential equation.

6.1   TRAVELING SALESMAN

Chapter 5 presented several methods to modify the crossover and mutation operators in order for a genetic algorithm to tackle reordering or permutation problems. It's time to try this brand of genetic algorithm on the famous traveling salesman problem, which represents a classic optimization problem that cannot be solved using traditional techniques [although it has been successfully attacked with simulated annealing (Kirkpatrick et al., 1983)]. The goal is to find the shortest route for a salesman to take in visiting N cities. This type of problem appears in many forms, with some engineering applications that include the optimal layout of a gas pipeline, design of an antenna feed system, configuration of transistors on a very large-scale integration (VLSI) circuit, or sorting objects to match a particular configuration. Euler introduced a form of the traveling salesman problem in 1759, and it was formally named and introduced by the Rand Corporation in 1948 (Michalewicz, 1992).

The cost function for the simplest form of the problem is just the distance traveled by the salesman for the given ordering (x_n, y_n), $n = 1, \ldots, N$ given by

$$cost = \sum_{n=0}^{N} \sqrt{(x_n - x_{n+1})^2 + (y_n - y_{n+1})^2} \tag{6.1}$$

where (x_n, y_n) are the coordinates of the nth city visited. For our example, let's put the starting and ending point at the origin, so $(x_0, y_0) = (x_{N+1}, y_{N+1}) = (0, 0) = $ starting and ending point. This requirement ties the hands of the algorithm somewhat. Letting the starting city float provides more possibilities of optimal solutions.

To treat such a problem, we must first define an alphabet. Any alphabet will do, of course, and, in fact, binary alphabets have been used for this problem, but it may make sense to define the alphabet as integers ranging from 1 through N. The crossover operator is a variation of the cycle crossover (CX) described in Chapter 5. Here, however, we randomly select a location in the chromosome where the integers are exchanged between the two parents. Unless the exchanged integers are the same, each offspring has a duplicate integer. Next, the repeated integer in offspring 1 is switched with the integer at that site in offspring 2. Now a different integer is duplicated, so the process iterates until we return to the first exchanged site. At this point, each offspring contains exactly one copy of each integer from 1 to N. The mutation operator randomly chooses a string, selecting two random sites within that string, and exchanges the integers at those sites.

We'll initially look at this problem with $N = 13$ cities. Given the fixed starting and ending points, there are a total of $13!/2 = 3.1135 \times 10^9$ possible combinations to check. To test the algorithm, we will start with a configuration where all the cities lie in a rectangle as shown in Figure 6.1. We know that the minimum distance is 14. The genetic algorithm parameters for this case are $N_{ipop} = 800$, $N_{pop} = 400$, $N_{good} = 200$, and $\mu = 0.04$. The algorithm found the solution in 35 generations as shown in Figure 6.2.

Now let's try a more difficult configuration. Randomly placing the 13 cities in a $100 \text{ km} \times 100 \text{ km}$ square doesn't have an obvious minimum path. How do we know that the genetic algorithm has arrived at the solution? We don't. We used two ways to check the algorithm. First, the algorithm had $N_{ipop} = 800$, $N_{pop} = 400$, $N_{good} = 200$, and $\mu = 0.04$. This algorithm was run five times over 200 generations with the best minimum results shown in Table 6.1. The minimum cost for two of the runs was 316.9 with the other three runs very close behind. The second way to check the result is to make a run with a larger population. Run 6 had $N_{ipop} = 1600$, $N_{pop} = 800$, $N_{good} = 400$, and $\mu = 0.04$ and arrived at the same minimum result as two of the five smaller runs. After these two checks, how sure are we that this is *the* global minimum? Well, we're kind of sure... but wouldn't bet our lives on it! One final check—there is a formula for the expected length of a traveling salesman path with random placement of the cities (Michalewicz, 1992):

$$cost = 0.765\sqrt{(N + 1)A} \tag{6.2}$$

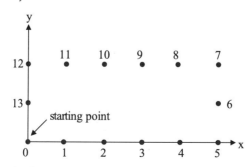

Figure 6.1 Graph of 13 cities arranged in a rectangle. The salesman starts at the origin and visits all 13 cities once and returns to the starting point. The obvious solution is to trace the rectangle which has a distance of 14.

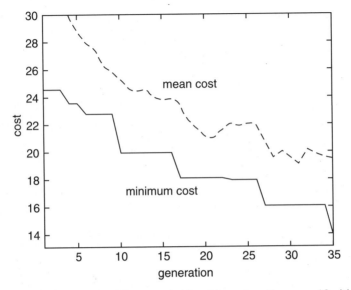

Figure 6.2 Convergence of the genetic algorithm when there are 13 cities on a rectangle as shown in Figure 6.10.

where A is the area that contains the randomly located cities ($A = 100 \times 100$). Thus the estimate for the cost of our random 13 cities is $cost = 0.765\sqrt{(13 + 1)(100 \times 100)} = 286.24$. Our results are pretty close (from Table 6.1 the shortest path $\simeq 317$), so our confidence of a correct answer is boosted.

Increasing the number of cities in the rectangle to 19 raises the number of combinations to $19!/2 = 6.0823 \times 10^{16}$. Adding a small number of additional cities has certainly increased the complexity of the problem. When the cities are arranged in a rectangle, we know the answer is 20.

TABLE 6.1 Best Minimum Results of the Cost of Sales Runs*

Run	Minimum Cost ($)
1	318.0073
2	317.7250
3	316.8982
4	316.8982
5	318.0073
Average	317.5072
6	316.8982

*The first five runs were done with half the population of the sixth run.

TABLE 6.2   Results for the 19-City Traveling Salesman Problem (Where the Exact Answer is 20)

Run	N_{ipop}	N_{pop}	N_{good}	μ	# Generations	Min Cost
1	1000	500	250	0.04	200	24.0645
2	1200	600	300	0.04	200	21.4142
3	1400	700	350	0.04	200	21.4142
4	1600	800	400	0.04	200	23.8416

Table 6.2 provides a few results. We didn't do much experimenting or optimizing of the code to get these results. Other people have achieved stunning results for 30 to 512 cities (Whitley et al., 1991; Michalewicz, 1992).

6.2   LOCATING AN EMERGENCY RESPONSE UNIT REVISITED

Finding the location of an emergency response unit described in Chapter 4 had a cost surface with two minima. Running the continuous and binary genetic algorithms revealed that the continuous genetic algorithm was superior. One of the problems with the binary genetic algorithm is the use of binary numbers to represent parameter values. In this chapter, we solve the same problem with a binary genetic algorithm, but use a Gray code to represent the parameters.

Gray codes don't always improve the convergence of a genetic algorithm. The convergence graph in Figure 6.3 shows that the Gray code did make a difference in this instance. Implementing the Gray code in the genetic algorithm slows down the algorithm, because the translation of the binary code into binary numbers is time-consuming. We're somewhat skeptical of adding the Gray code translation to our genetic algorithms, so we usually don't. The result here shows that a small improvement is possible with the Gray code.

6.3   DECODING A SECRET MESSAGE

This example uses a real parameter genetic algorithm to break a secret code. A message consisting of letters and spaces is encoded by randomly changing one letter to another letter. For instance, all d's may be changed to c's and spaces changed to q's. If the message uses every letter in the alphabet, then there are a total of 27! possible codes, with only one being

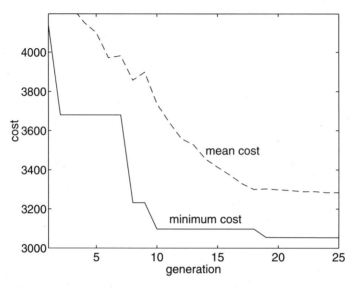

Figure 6.3 Convergence graph for the emergency response unit problem from Chapter 4 when a Gray code is used.

correct. If the message uses S symbols, then there are $27! - S!$ possible encodings that work.

A chromosome consists of 27 genes with unique values from 1 to 27. A 1 corresponds to a space and 2 through 27 correspond to the letters of the alphabet. Letters and spaces in the *message* receive the appropriate numeric values. The cost is calculated by subtracting the guess of the message from the known message, taking the absolute value, and summing:

$$cost = \sum_{n=1}^{N} |message(n) - guess(n)| \qquad (6.3)$$

We know the message when the cost is zero.

As an example, let's see how long it takes the genetic algorithm to find the encoding for the message "bonny and amy are our children." This message has 30 total symbols, of which 15 are distinct. Thus, 15 of the letters must be in the proper order, while the remaining 12 letters can be in any order. The genetic algorithm used the following constants: $N_{ipop} = 800$, $N_{pop} = 400$, $N_{good} = 40$, and $\mu = 0.02$. It found the message in 68 generations as shown in Figure 6.4. Progress on the decoding is shown in Table 6.3.

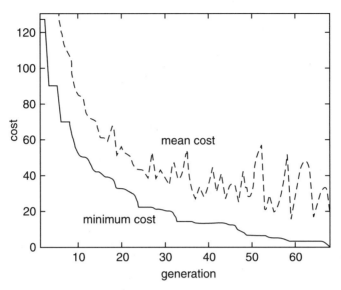

Figure 6.4 The genetic algorithm decodes the message "bonny and amy are our children" in 68 generations.

A more difficult message is "jake can go out with my beautiful pets and quickly drive to see you." This message lacks only x and z. It has 25 distinct symbols and a total of 65 total symbols. Figure 6.5 shows the convergence in this case with $N_{ipop} = 1000$, $N_{pop} = 500$, $N_{good} = 40$, and $\mu = 0.02$. The algorithm found the solution in 86 generations. Progress on the decoding is shown in Table 6.4.

TABLE 6.3   Progress of the Genetic Algorithm as it Decodes the Secret Message

Generation	Message
1	amiizbditbdxzbdqfbmvqbeoystqfi
10	krooy aoe any aqf rwq gbpseqfo
20	crooy aoe any aqf rwq gdiheqfo
30	dpooy aoe any arf pwr ghikerfo
40	bqmmz amd anz are qur cfildrem
50	bonnz and amz are osr cghldren
60	bonny and ajy are our children
68	bonny and amy are our children

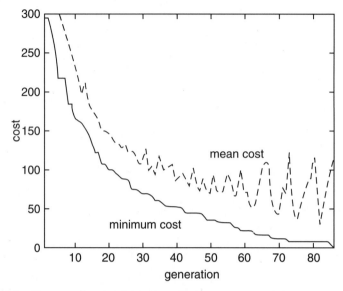

Figure 6.5 The genetic algorithm decodes the message "jake can go out with my beautiful pets and quickly drive to see you" in 86 generations.

6.4 ROBOT TRAJECTORY PLANNING

Robots imitate biological movement, and genetic algorithms imitate biological survival. The two topics seem to be a perfect match, and many researchers have made that connection. Several studies have investigated the use of genetic algorithms for robot trajectory planning (Davidor, 1991;

TABLE 6.4 **Progress of the Genetic Algorithm as it Decodes the Secret Message**

Generation	Message
1	vhte fhb po olq zjqk ds mehlqjxlu neqr hbg wljftus gcjae qo ree sol
10	cahd bas np pxt iqtf kz edaxtqwxj vdtl asg oxqbhjz grqud tp ldd zpx
20	jakh dar go out wftb mx nhautfcui phty are sufdkix ezfqh to yhh xou
30	faje can gp pvs yish mx reavsikvl ueso and qvicjlx dwize sp oee xpv
40	kaje can dp pvt yitg mx reavtifvl oets anb qvicjlx bwize tp see xpv
50	jake can dp pvt xitg my heavtifvl oets anb qvickly bwize tp see ypv
60	jake can gp put xith my deautiful oets anb quickly bvize tp see ypu
70	jake can go out xith my beautiful pets and quickly dwize to see you
80	jake can go out xith my beautiful pets and quickly dwive to see you
86	jake can go out with my beautiful pets and quickly drive to see you

Davis, 1991; Pack, 1996). For example, the goal is to move a robot arm in an efficient manner while avoiding obstacles and impossible motions. The even more complicated scenario of moving the robot arm when obstacles are in motion has been implemented with a parallel version of a genetic algorithm (Chambers, 1995). Another application simulated two robots fighting. A genetic algorithm was used to evolve a robot's strategy to defeat its opponent (Yao, 1995).

A robot trajectory describes the position, orientation, velocity, and acceleration of each robot component as a function of time. In this example, the robot is a two-link arm having two degrees of freedom in a plane called the robot workspace (Figure 6.6) (Pack, 1996). For calculation purposes, this arm is approximated by two line segments in Cartesian coordinates as shown in Figure 6.7. Each joint has its own local coordinate system that can be related to the base x_0, y_0 coordinate system (located at the shoulder joint). The end-effector or tip of the robot arm is of most interest and has a local coordinate system defined by x_2, y_2. An intermediate coordinate system at the elbow joint is defined by x_1, y_1. Using the Donauit-Hartenberg parameters, one can transform an end-effector position in terms of the x_0, y_0 coordinates by

$$
\begin{bmatrix}
\cos\theta_{12} & -\sin\theta_{12} & 0 & \ell_1\cos\theta_1 + \ell_2\cos\theta_{12} \\
\sin\theta_{12} & \cos\theta_{12} & 0 & \ell_1\sin\theta_1 + \ell_2\sin\theta_{12} \\
0 & 0 & 1 & 0 \\
0 & 0 & 0 & 1
\end{bmatrix}
\begin{bmatrix}
x_2 \\
y_2 \\
z_2 \\
1
\end{bmatrix}
=
\begin{bmatrix}
x_0 \\
y_0 \\
z_0 \\
1
\end{bmatrix}
\quad (6.4)
$$

where

$x_2, y_2, z_2 =$ position of end-effector with respect to coordinate system 2 (end-effector based coordinate system)

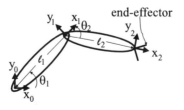

Figure 6.6 Diagram of a two-dimensional robot arm with two links. Link 1 pivots about coordinate system 0 and link 2 pivots about coordinate system 1. Coordinate system 3 has an origin at the end-effector.

Figure 6.7   The robot arm in Figure 6.6 is more simply described by two line segments as shown here.

x_0, y_0, z_0 = position of end-effector with respect to the base coordinate
    system

$\cos \theta_{12}$ = $\cos \theta_1 \cos \theta_2 - \sin \theta_1 \sin \theta_2$

$\sin \theta_{12}$ = $\sin \theta_1 \cos \theta_2 + \cos \theta_1 \sin \theta_2$

ℓ_1 = length of link 1

ℓ_2 = length of link 2

θ_1 = angle between x_0 axis and link 1

θ_2 = angle between x_1 axis and link 2

Thus, knowing the length of the links and the angles allows us to transform any points on the robot arm from the x_2, y_2 coordinate system to the x_0, y_0 coordinate system. Our goal is to find the optimal path for the robot to move through its environment without colliding with any obstacles in the robot workspace.

Although following the end-effector path through Cartesian space (x_0 and y_0 axes) is easiest to visualize, it is not of the most practical value for optimization. First, the calculation of joint angles at each point along the path is difficult. Second, the calculations can encounter singularities that are difficult to avoid. An alternative approach is to formulate the trajectory problem in the configuration space (θ_1 and θ_2 axes) that governs the position of the end-effector. Although numerically easier, it can result in complicated end-effector paths. We will go with the numerically easier version and let the genetic algorithm optimize in configuration space for this example.

Obstacles in the form of impossible robot joint angle combinations must be taken into account when designing the cost function. It can be shown that point obstacles are contained within an elliptical region in configuration space (Pack, 1996). As an example, a point obstacle in the world space transforms into a curved line in configuration space (Figure 6.8) (Pack, 1996). This line is nicely contained within an ellipse, and an ellipse is much easier to model as an obstacle.

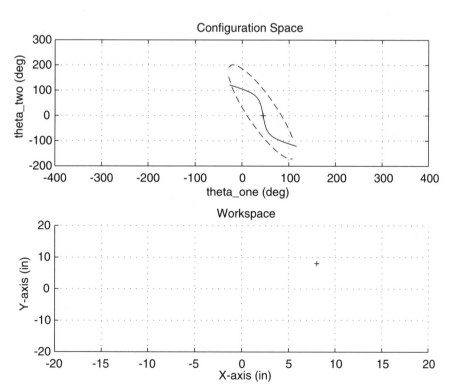

Figure 6.8   The point obstacle in the lower graph transforms into a curved line in configuration space as shown in the upper graph. This curved line is contained within an elliptical region denoted by the dashed line. In configuration space, this ellipse forms a boundary that the robot arm cannot pass through.

The cost function is merely the length of the line needed to get from the starting point to the ending point in the configuration space. Rather than attempt to find a continuous path between the start and destination points, piecewise line segments are used. This example establishes a set number of line segments before the genetic algorithm begins. Consequently, the length of all the chromosomes are the same. Others have used variable length chromosomes to find the optimum path. If the reader is interested in this approach, please refer to (Davidor, 1991).

The first example has four obstacles in the configuration space with start and stop points in obscure parts of the space. Only three intermediate points or four line segments are permitted to complete the shortest path from the start to the finish. The binary genetic algorithm had $N_{ipop} = N_{pop} = 80$ members in the population and ran for 10 generations. The first generation had an optimal path length of 11.06 units, as shown in Figure 6.9. After 10 generations the minimum cost reduced to 9.656 units, and its path in configuration space is shown in Figure 6.10. Adding more intermediate points would give the algorithm more freedom to find a better solution.

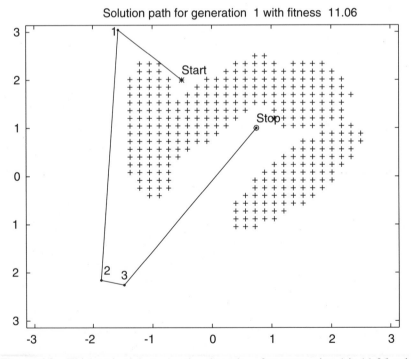

Figure 6.9 The best path between the obstacles after generation 1 is 11.06 units long.

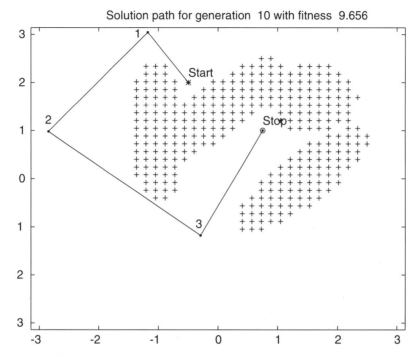

Figure 6.10   The best path between the obstacles after generation 10 is 9.656 units long.

A second example begins with a real world problem with five-point obstacles in world space that transformed into an ellipse in the configuration space. Again, the binary genetic algorithm had $N_{ipop} = N_{pop} = 80$ members in the population and ran for 10 generations. The path after the first generation is shown in Figure 6.11 and has a cost of 7.321 units. After 10 generations the minimum cost reduced to 6.43 units, and its path in configuration space is shown in Figure 6.12. This optimal solution translates back to world space, as shown in Figure 6.13, where the symbols $*$ and $\bullet$ denote the starting and ending robot end-effector positions, respectively. The elliptical obstacle shapes in Figures 6.12 and 6.13 translate into points (denoted by $+$ signs) in Figure 6.13.

6.5 STEALTH DESIGN

A stealth airplane is difficult to detect with conventional radar. Engineers use a combination of materials and shaping to reduce the radar cross section of an airplane. This example shows how the radar cross section of a simple

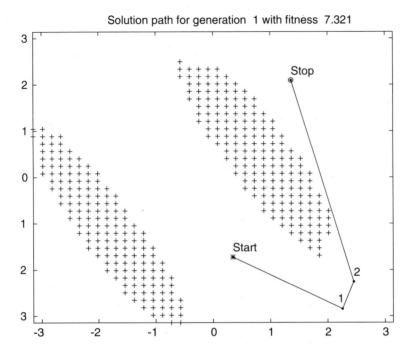

Figure 6.11 The best path between the obstacles after generation 1 has a length of 7.321 units.

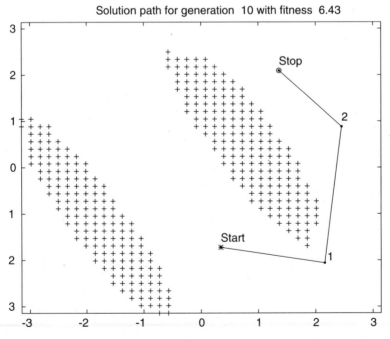

Figure 6.12 The best path between the obstacles after generation 10 has a length of 6.43 units.

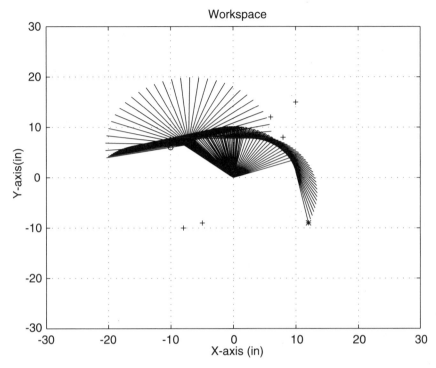

Figure 6.13   This is the actual movement of the robot arm through the obstacles in world space (denoted by + signs). The plus signs transform into the elliptical regions shown in configuration space (Figures 6.7 and 6.8).

two-dimensional reflector can be modified by the placement of absorbing materials next to it. This type of reflector design is also of interest to satellite antenna manufacturers to lower sidelobe levels and reduce the possibility of interference with the desired signal.

This example demonstrates how to use genetic algorithms to find resistive loads that produce the lowest maximum backscatter relative sidelobe level from a perfectly conducting strip. The backscattering pattern of a 6λ strip appears in Figure 6.14, and its relative sidelobe level is about 13.33 dB below the peak of the main beam. The radar cross section is given in terms of dBlambda or decibels above one wavelength. A model of the loaded strip is shown in Figure 6.15. Assuming the incident electric field is parallel to the edge of the strip, the physical optics backscattering radar cross section is given by (Haupt, 1995).

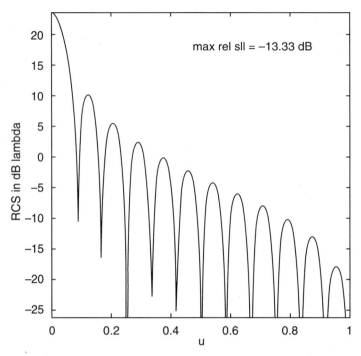

Figure 6.14   Radar cross section of a 6λ-wide perfectly conducting strip.

$$\sigma(\phi) = \frac{k}{4} \left| 4asSa(2kau\phi) + \sum_{n=1}^{N} \left(\frac{2b_n s}{0.5 + \eta_n s} \right) \right.$$

$$\left. \times Sa(kb_n u) \cos \left[2k \left(a + \sum_{m=1}^{n-1} b_m + \frac{b_n}{2} \right) u \right] \right|^2$$

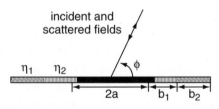

Figure 6.15   Diagram of a perfectly conducting strip with symmetric resistive loads placed at its edges.

where

$$s = \sin \phi$$
$$u = \cos \phi$$
$$2a = \text{width of perfectly conducting strip}$$
$$b_n = \text{width of load } n = \sum_{m=1}^{B_w} b_w[m]2^{1-m}W$$
$$\eta_n = \text{resistivity of load } n = \sum_{m=1}^{B_r} b_r[m]2^{1-m}R$$
$$Sa = \frac{\sin x}{x}$$

$B_w, B_r = $ number of bits representing the strip width and resistivity

$b_w, b_r = $ array of binary digits that encode the values for the strip widths and resistivities.

$W, R = $ width and resistivity of the largest quantization bit

Eight resistive loads are placed on each side of a perfectly conducting strip that is 6λ wide. The widths and resistivities of these loads are optimized to reduce the maximum relative sidelobe level. Both the width and resistivity of each load are represented by 5 quantization bits, and $W = 1$ and $R = 5$. The optimized values arrived at by genetic algorithms are

$$\eta_n = 0.16, 0.31, 0.78, 1.41, 1.88, 3.13, 4.53, 4.22$$
$$w_n = 1.31, 1.56, 1.94, 0.88, 0.81, 0.69, 1.00, 0.63\lambda$$

These values result in a maximum relative backscattering sidelobe level of -33.98 dB. Figure 6.16 shows the optimized backscattering pattern. The peak of the mainbeam is about 6 dB higher than the peak of the mainbeam of the 6λ perfectly conducting strip backscattering pattern in Figure 6.14. In exchange for the increase in the mainbeam, the peak sidelobe level is 15 dB less than the peak sidelobe level in Figure 6.14. In other words, compared to the 6λ perfectly conducting strip, this object is easier to detect by a radar looking at it from the broadside, but it is more difficult to detect looking off broadside.

The resistive loads attached to the perfectly conducting strip were also optimized using a quasi-Newton method that updates the Hessian matrix using the Broyden–Fletcher–Goldgarb–Shanno (BFGS) formula. A true gradient search was not used because the derivative of equation (6.5) is difficult to calculate. The quasi-Newton algorithm performed better than genetic algorithms for 10 or less loads. Using the quasi-Newton method in the previous example resulted in a maximum relative sidelobe level of -36.86 dB. When 15 loads were optimized, genetic algorithms were clearly superior.

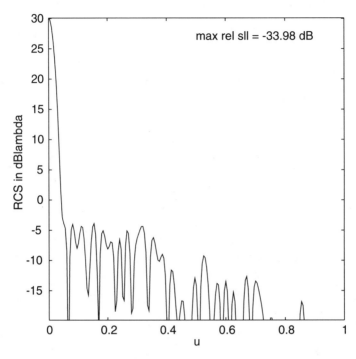

Figure 6.16 This is the radar cross section of the 6λ-wide strip with eight resistive loads placed at its edges. The maximum sidelobe level is -33.98 dB less than and the peak of the mainbeam, which is a 20.78 dB reduction.

6.6 BUILDING A DYNAMICAL INVERSE MODEL

Sometimes, we have collected large amounts of data but have not developed adequate theories to explain the data. Other times, the theoretical models are so complex that it is extremely computer intensive to use them. In either case, it is sometimes useful to begin with the available data to define parameters of a stochastic model that minimizes some mathematical normed quantity, that is, a cost. Our motivation here lies in trying to predict environmental variables. Since environmental prediction is a highly nonlinear problem, it is not easy to solve. In recent years, many scientists have been using the theory of Markov processes combined with a least squares minimization technique to build stochastic models of environmental variables in atmospheric and oceanic science (Hasselmann, 1976; Penland, 1989; Penland and Ghil, 1993). One example is trying to predict the time evolution of sea surface temperatures in the western Pacific Ocean as a model of the rises and falls of the El Niño/Southern Oscillation (ENSO) cycle. This problem has proven challenging. However, stochastic models have performed as well as the dynamical ones in predicting future ENSO cycles (Penland and Magorian 1993; Penland, 1996). Another application involves finding ways to simplify climate prediction. We now build very complex climate models that require huge amounts of computer time to run. There are occasions when it would be useful to predict the stochastic behavior of just a few of the key variables in a large atmospheric model without concern for the details of day-to-day weather. One such application is when an atmospheric climate model is coupled to an ocean model. Since the time scale of change of the atmosphere is so much faster than that of the ocean, its scale dictates the Courant-Friedichs-Levy criteria, which limits the size of the allowable time step. For some problems, it would be convenient to have a simple stochastic model of the atmosphere to use in forcing an ocean model. Recent attempts have shown that such models are possible and perhaps useful for computing responses to forcing (Branstator and Haupt, 1997a, b). However, the least squares techniques typically used to build these models assume a Markov process. This assumption is not valid for most environmental time series. Would a different method of minimizing the function produce a better match to the environmental time series? This is an interesting question without a clear answer. Before answering it using large climate models, it is convenient to begin with simple low-dimensional models of analytical curves.

We wish to use a genetic algorithm to compute parameters of a model of a simple curve that is clearly non-Markovian. In particular, we wish to fit a model:

$$\mathbf{x_t} = \mathbf{Ax} \tag{6.5}$$

to a time series of data. Here, $\mathbf{x}$ is an N-dimensional vector; $\mathbf{x_t}$ its time tendency; and $\mathbf{A}$ is an $N \times N$ matrix relating the two. Note that most first-order time-dependent differential equations can be discretized to this form. Our goal is to find the matrix $\mathbf{A}$ that minimizes the cost

$$cost = \left\langle (\mathbf{x_t} - \mathbf{Ax})^P \right\rangle \tag{6.6}$$

where P is any appropriate power norm that we choose. The least squares methods use $P = 2$, or, an L^2 norm. The angular brackets denote a sum over all of the data in the time series.

The time evolution of the spiral curve we will model appears in Figure 6.17. This curve was generated as $(X, Y, Z) = (\sin(t), \cos(t), t)$, with $t = [0, 10\pi]$ in increments of $\pi/50$. Note that for this problem, computa-

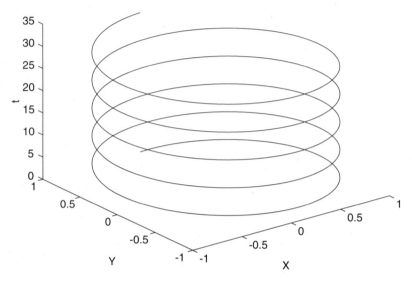

Figure 6.17   The spiral curve specified by $(X, Y, Z) = (\cos(t), \sin(t), t)$ for $t = [1, 10\pi]$.

tion of the cost function becomes intensive, since the cost function involves a summation over 500 time increments. However, even with the reasonably large population size and number of generations (70) that we computed, the computer time required was not excessive. A continuous parameter genetic algorithm is applied to this curve with an initial population size of $N_{ipop} = 400$, population size of $N_{pop} = 100$, and a mutation rate of $\mu = 0.2$. Since the genetic algorithm is oblivious to which value of P we choose, we experimented a bit and found the best results for moderate P. The solution displayed here uses $P = 4$. Evolution of the minimum cost appears in Figure 6.18. We notice that the cost decreases several orders of magnitude over the 70 generations. The result appears in Figure 6.19. We see that the general shape of the spiral curve is captured rather well. The bounds in X and Y are approximately correct, but the evolution in $Z = t$ is too slow. We found that this aspect of our model was rather difficult to capture. In terms of dynamical systems, we were able to find the attractor, but not able to exactly model the evolution along it. For comparison, a standard least squares technique is used to solve the same problem. The result appears as Figure 6.20. We can see that the least squares method could not even come close to capturing the shape of the attractor. Of course,

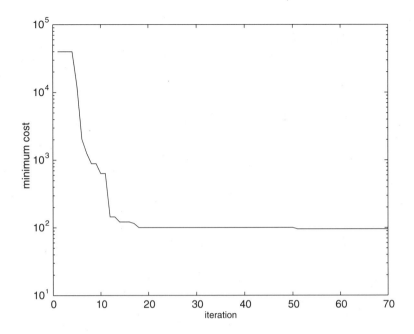

Figure 6.18 Evolution of the mimimum cost of the genetic algorithm which produces a dynamical inverse model of the spiral curve in Figure 6.17.

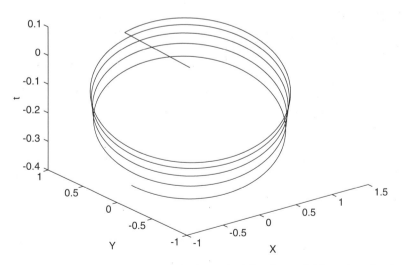

Figure 6.19 The genetic algorithm's dynamical fit of a model based on the time series of the spiral curve in Figure 6.17.

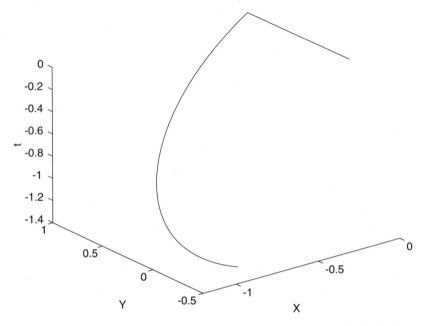

Figure 6.20 A linear least square dynamical fit of a model based on the time series of the spiral curve in Figure 6.17.

we can fine tune the least squares method by adding a noise term in the cost function. We can do that for the genetic algorithm (GA) as well. The advantage of the GA is that it is quite simple to add complexity to the cost function. For this simple model, doing that adds nothing to the solution of the problem since it can be completely specified with the nine degrees of freedom in the matrix.

This is only preliminary work on producing dynamical inverse models using a GA. There are many interesting models to try. It is unclear how far we can go with this technique.

6.7 SOLVING HIGH-ORDER NONLINEAR PARTIAL DIFFERENTIAL EQUATIONS

Two of the tools of scientists and engineers are differential and partial differential equations. Normally, we don't think of these equations as minimization problems. However, if we want to find for which values a differential equation is zero (a form in which we can always cast the system), we can look for the minimum of its absolute value. Koza (1992) demonstrated that a genetic algorithm could solve a simple differential equation by minimizing the value of the solution at 200 points. To do this, he numerically differentiated at each point and fit the appropriate solution using a GA. Here, we are interested in demonstrating that a genetic algorithm is a useful technique for solving a highly nonlinear differential equation that is formally nonintegrable. However, we do know the solitary wave approximate solutions. Solitary waves, or solitons, are permanent-form waves for which the nonlinearity balances the dispersion to produce a coherent structure. The equation that we will solve is the Super Korteweg-deVries equation (SKDV), a fifth-order nonlinear partial differential equation:

$$u_t + \alpha u u_x + \mu u_{xxx} - \nu u_{xxxxx} = 0 \tag{6.7}$$

The functional form is denoted by u; time derivative by the t subscript; spatial derivative by the x subscript; and α, μ, and ν are parameters of the problem. We wish to solve for waves that are steadily translating, so we write the t variation using a Galilean tranformation, $X = x - ct$, where c is the phase speed of the wave. Thus, our SKDV becomes a fifth-order, nonlinear ordinary differential equation:

$$(\alpha u - c)u_X + \mu u_{XXX} - \nu u_{XXXXX} = 0 \tag{6.8}$$

Boyd (1986) extensively studied methods of solving this equation. He expanded the solution in terms of Fourier series to find periodic cnoidal wave solutions (solitons that are repeated periodically). Among the methods used are the analytical Stokes' expansion, which intrinsically assumes small amplitude waves, and the numerical Newton-Kantorovich iterative method, which can go beyond the small amplitude regime if care is taken to provide a very good first guess. Haupt and Boyd (1988a) were able to extend these methods to deal with resonance conditions. These methods were generalized to two dimensions to find double cnoidal waves (two waves of differing wave number on each period) for the integrable Korteweg de Vries equation (1988b) and the nonintegrable Regularized Long Wave Equation (Haupt, 1988). However, these methods require careful analytics and programming that is very problem specific. Here, we are able to add a simple modification to the cost function of our genetic algorithm to obtain a similar result.

To find the solution of equation (6.8), we expand the function u in terms of a Fourier cosine series:

$$u(X) \simeq u_K(X) = \sum_{k=1}^{K} a_k \cos(kx) \tag{6.9}$$

Without loss of generality, we have assumed that the function is symmetric about the X-axis by not including sine functions. In addition, we use the "cnoidal convention" by assuming that the constant is 0. Now, we can easily take derivatives as powers of the wave numbers to write the cost that we wish to minimize as:

$$cost(u_K) = \sum_{k=1}^{K} \left[-k(\alpha u - c) + k^3 \mu + k^5 \nu \right] a_k \sin(kx) \qquad (6.10)$$

This is reasonably easy to put into the cost function of a genetic algorithm where we want to find the coefficients of the series, a_k. The only minor complication is computing u to insert into the cost function, equation (6.10). However, this is merely one extra line of code.

The parameters that we used here are $\nu = 1$, $\mu = 0$, $\alpha = 1$, and a phase speed of $c = 14.683$ to match with a well known nonlinear solution. Note that the phase speed and amplitude of solitary-type waves are interdependent. We could instead have specified the amplitude and

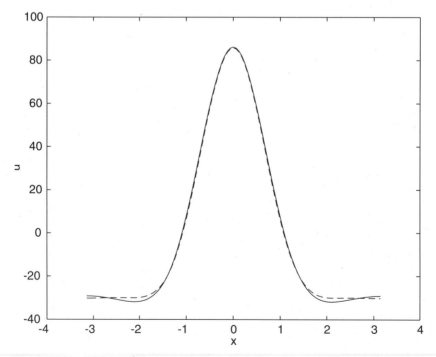

Figure 6.21 The cnoidal wave of the Super Korteweg de Vries equation. Solid line—exact solution. Dashed line—the genetic algorithm solution.

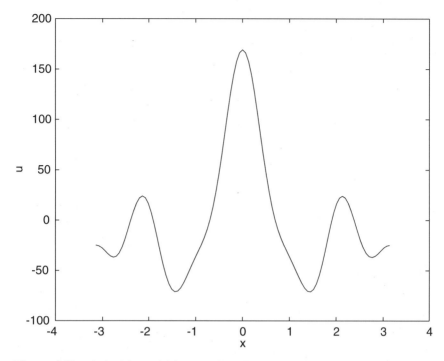

Figure 6.22   A double cnoidal wave of the Super Korteweg de Vries equation as computed by the genetic algorithm.

solved for the phase speed. It is equivalent. We computed the coefficients, a_k, to find the best cnoidal wave solution for $K = 6$. We used $N_{ipop} = 500$, $N_{pop} = 100$, $\mu = 0.2$, and 70 iterations. We evaluated the cost function at merely two points for this run and summed their absolute value. The results appear in Figure 6.21. The solid line is the "exact" solution reported by Boyd (1986) and the dashed line is the genetic algorithm's approximation to it. They are barely distinguishable. For interest's sake we show a genetic algorithm solution that converged to a double cnoidal wave as Figure 6.22.

So we see that genetic algorithms show promise for finding solutions of differential and partial differential equations, even when these equations are highly nonlinear and have high-order derivatives.

BIBLIOGRAPHY

Boyd, J. P., 1986, "Solitons from sine waves: analytical and numerical methods for non-integrable solitary and cnoidal waves," *Physica*, **21D**, pp. 227–246.

Branstator, G., and S. E. Haupt, 1997a, "An empirical model of barotropic atmospheric dynamics and its response to forcing," submitted to *J. Climate.*

Branstator, G., and S. E. Haupt, 1997b, "A sensitivity analysis of an empirical climate model," to be submitted to *J. Atmos. Sci.*

Chambers, L., (Ed.) 1995, *Genetic Algorithms, Applications*, Vol. I, New York: CRC Press.

Davidor, Y., 1991, *Genetic Algorithms and Robotics*, River Edge, N.J.: World Scientific.

Davis, L., 1991, *Handbook of Genetic Algorithms*, New York: Van Nostrand Reinhold.

Hasselmann, K., 1976, "Stochastic climate models. Part I: Theory," *Tellus*, **28**, pp. 473–485.

Haupt, R. L., 1995 April, "An introduction to genetic algorithms for electromagnetics," *IEEE Antennas Propagat. Mag.*, **37**, (2), pp. 7–15.

Haupt, S. E., Ph.D. 1988, *Solving Nonlinear Wave Problems with Spectral Boundary Value Techniques*, Dissertation, University of Michigan, Ann Arbor, 157 pp.

Haupt, S. E., and J. P. Boyd, 1988a, "Modeling nonlinear resonance: A modification to the Stokes' perturbation expansion," *Wave Motion*, **10**, pp. 83–98.

Haupt, S. E., and J. P. Boyd, 1988b, "Double cnoidal waves of the Korteweg De Vries equation: solution by the spectral boundary value approach," *Physica D*, **50**, pp. 117–134.

Holland, J. H., 1992 July, "Genetic algorithms," *Sci. Amer.*, pp. 66–72.

Kirkpatrick, S., C. D. Gelatt, and M. P. Vecchi, 1983, "Optimization by simulated annealing," *Science*, **220**, pp. 671–680.

Koza, J. R., 1992, "The Genetic Programming Paradigm: Genetically Breeding Populations of Computer Programs to Solve Problems," in B. Soucek (Ed.), *Dynamic, Genetic, and Chaotic Programming, The Sixth Generation*, New York: John Wiley, pp. 203–321.

Michalewicz, Z., 1992, *Genetic Algorithms + Data Structures = Evolution Programs*, New York: Springer-Verlag.

Pack, D., G. Toussaint, and R. Haupt, 1996, "Robot trajectory planning using a genetic algorithm," International Symposium on Optical Science, Engineering, and Instrumentation, SPIE's Annual Meeting, Denver, CO.

Penland, C., 1989, "Random forcing and forecasting using principal oscillation pattern analysis," *Mon. Weather Rev.* **117**, pp. 2165–2185.

Penland, C., 1996, "A stochasic model of IndoPacific sea surface temperature anomalies," *Physica D*, **98**, pp. 534–558.

Penland, C., and M. Ghil, 1993, "Forecasting northern hemisphere 700mb geopotential height anomalies using empirical normal modes," *Mon. Weather Rev.* **121**, p. 2355.

Penland, C., and T. Magorian, 1993, "Prediction of NINO3 sea-surface temperatures using linear inverse modeling," *J. Climate*, **6**, p. 1067.

Whitley, D., T. Starkweather, and D. Shaner, 1991, "The traveling salesman and sequence scheduling: Quality solutions using genetic edge recombination," in L. Davis, (Ed.), *Handbook of Genetic Algorithms*, New York: Van Nostrand Reinhold.

Yao, X., (Ed.), 1995, *Progress in Evolutionary Computation*, Springer-Verlag, New York.

CHAPTER 7

EVOLUTIONARY TRENDS

Evolutionary modeling in general and genetic algorithms in particular have had a relatively short history. However, it is a history that progressed rapidly. In this chapter, we mention a few of the pioneers of genetic research, then briefly summarize some current research areas, before making just a few comments on what the future may hold.

7.1 THE PAST

Modeling biological evolution on a computer began in the 1960s. Rechenberg (1965) introduced evolution strategies in Europe. His first versions of the algorithms used real-valued parameters and began with a parent and a mutated version of a parent. Whichever had the highest cost was discarded. The winner produced a mutated version and the process repeated. Populations and crossover were not incorporated until later years. A general version of the algorithm, known as the $\mu + \lambda$ evolution strategy, was developed (Schwefel, 1995). In this strategy, μ parents produce λ offspring. In succeeding generations, only μ of the λ offspring are allowed to become parents of the next generation. None of the parents from the previous generation are allowed to produce offspring again. Fogel introduced the concept of evolutionary programming in the United States at about the same time as Rechenberg's work (Fogel et al., 1966). Today this area has become an exciting attempt to have computers write their own codes.

Genetic algorithms came on the scene at the same time as well. Holland's original work was summarized in his book. He was the first to try to develop a theoretical basis for genetic algorithms through his schema theorem. The work of De Jong (1975) showed the usefulness of the genetic algorithm for function optimization and made the first concerted effort to find optimized genetic algorithm parameters. Goldberg has probably contributed to the most fuel to the genetic algorithm fire with his successful applications and excellent book (1989). Since then, many versions of evolutionary programming have been tried with varying degrees of success.

7.2 THE PRESENT

Research with genetic algorithms is a blooming field. Keeping track of all the research has become an impossible task. Virtually all areas of study have some application of genetic algorithms, including fields ranging from stock market predictions and portfolio planning to biochemistry and signal processing. Genetic algorithms have come out of the closet. They once were relegated to computer science symposia and theoretical testing of assorted aspects of this new algorithm. Today we find genetic algorithms solving problems of everyday interest. Much of the most recent and interesting research is not yet available in the peer-reviewed literature. That is why so many conference proceedings are referenced here. Our feeble attempt to summarize some of the many research areas follows. In addition we list some of the reservoir of information available to genetic algorithm novices and experienced users.

7.2.1 Other Research Areas

Chapters 4 and 6 presented several detailed examples of genetic algorithms. These examples barely scratch the surface of what has been done, let alone what could be done with genetic algorithms.

7.2.1.1 Radar and Communications Systems Genetic algorithms have found many applications in radar systems. A radar consists of a transmitter, a receiver, and an antenna (Figure 7.1). Research concentrates on the generation, transmission, propagation, scattering, and reception of the radar electromagnetic signal. Michielssen et al. (1993) used a genetic algorithm to synthesize a multilayer radar absorbing coating that maximizes

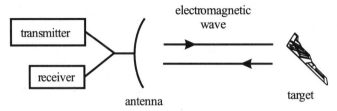

Figure 7.1 Diagram of a radar system.

absorption of an electromagnetic wave over a desired range of frequencies and incident angles. The optimization parameters were absorber thickness, permittivity, and permeability. Controlling the radar cross section of a metal reflector is of importance in stealth designs. Haupt (1995) investigated various methods of modifying the spacing and adding resistivity to optimally control electromagnetic scattering from a two-dimensional reflector. Li et al. used a genetic algorithm to optimize data storage of ultrawideband radar signatures. These signatures are correlated with radar returns to identify targets.

Genetic algorithms were also applied to the problem of thinning linear and planar arrays to obtain the lowest maximum relative sidelobe level over a specified bandwidth and scan angle. An overview of antenna array optimization with genetic algorithms is presented in Haupt (1994b). Other antenna optimizations include design of electrically loaded wire antennas (Boag et al., 1996), of yagi antennas (Linden and Altshuler, 1996), and of loaded monopole antennas (Altshuler and Linden, 1997). In addition, genetic algorithms have proven useful as an adaptive nulling algorithm (Haupt, 1997). An adaptive antenna places a null in the direction of an interference source in order to prevent the interference from being received by the antenna.

Antennas are of common interest to radar and communications systems. Another concern of communications systems is telecommunications network scheduling. A hybrid genetic algorithm that makes use of random search found optimal bandwidth packing in a given channel. In other words, as many messages as possible were placed on a given channel, taking into account the source, destination, requested start time, requested duration, bandwidth, and priority (Cox et al., 1991). Others have used genetic algorithms to optimize packet-switched communication networks (Carse et al., 1995) and optimize the assignment of radio frequencies to a military communications network (Kapsalis et al., 1995). These communication network problems use a version of a traveling salesman or scheduling genetic algorithm.

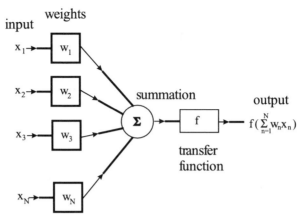

Figure 7.2   A single node in a neural network.

7.2.1.2   Neural Networks   Neural networks are an important branch of artificial intelligence (Haykin, 1994). These networks provide adaptive control of a process by modeling the nervous system of animals. Figure 7.2 shows a typical node in a neural network. The outputs from the nodes are weighted and fed into other nodes. Information transfers from node to node much as electrical signals pass from neuron to neuron in the body's nervous system. A neural network is partly characterized by the number of input ports it has and the number of layers of the nodes. Figure 7.3 shows a layer in a neural network. Input signals enter an initial layer. The output from this layer may pass to another layer, and so on until the final layer produces the output of the neural network. Weights, connectivity, number of nodes, and number of inputs are all parameters in neural network design.

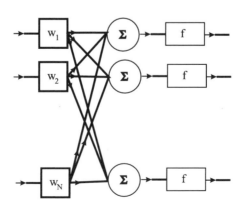

Figure 7.3   A layer in a neural network.

Genetic algorithms have been used to construct neural networks and optimize their complexity (Chambers, 1995; Harp et al., 1991; Miller et al., 1989). Reducing the number of nodes and layers and finding optimal connections between nodes are extremely difficult optimization problems. Once the network is built, it must be trained. Genetic algorithms have been used to train neural networks that distinguish between sonar returns from man-made objects and background noise (Porto et al., 1995). Genetic algorithms proved to be superior to traditional backpropagation. Another experiment showed that the genetic algorithm significantly outperformed backpropagation in training a neural network for predicting the optimum transistor width in a complementary metal-oxide semiconductor (CMOS) switch (Jansen and Frenzel, 1993).

7.2.1.3 *Signal Processing* Signal processing has made extensive use of genetic algorithms. An excellent review article appears in Tang et al. (1996), and we'll briefly summarize some of the research topics. Genetic algorithms have been applied to the synthesis of finite impulse-response (transfer function has all zeros) and infinite impulse-response (transfer function has zeros and poles) filters. Of special interest are adaptive filters that modify their coefficients to eliminate undesirable signals. Typically, gradient methods like the least mean square algorithm or gradient-lattice algorithm are used to adapt isobutylene-isoprene rubber (IIR) filter coefficients. Figure 7.4 shows a diagram of an adaptive IIR filter. An IIR filter has a multimodal cost surface that confuses gradient-based algorithms. Best results were obtained with a hybrid genetic algorithm that made use of a gradient search method (Ng et al., 1996). A flow chart of the genetic search learning algorithm for adaptive IIR filtering appears in Figure 7.5. Active

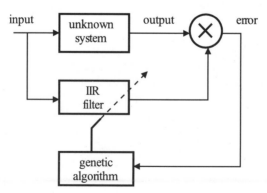

Figure 7.4 A diagram of an adaptive IIR filter for system identification.

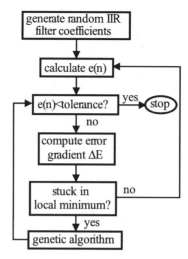

Figure 7.5   Flow chart of an adaptive IIR filter based on a genetic algorithm.

noise control is a technique that generates secondary sounds to cancel a primary sound. Headphones that cancel unwanted external sounds are an example of an application of this method. A genetic algorithm active noise control system has been proposed to attenuate noise (Tang et al., 1995) Genetic algorithms used in speech processing were found to outperform the commonly used dynamic time-warping technique (Tang et al., 1996).

7.2.1.4 Geophysics   Genetic algorithms (GAs) have also found use in geophysics. The most prevalent type of problem treated with a GA in the literature is the inverse problem. An inverse model defines parameters for a theoretical or numerical model based on experimental measurements. An example is determining the type of underground rock layers. Since it is not practical to take core samples of sufficient resolution to create good maps of the underground layers, modern techniques use seismic information or apply a current and measure the potential difference that gives a resistance. These various methods produce an underdetermined multimodal model of the earth. Fitting model parameters to match the data is regarded as a highly nonlinear process. Genetic algorithms have found recent success in finding realistic solutions for this inverse problem (Jervis and Stoffa, 1993; Jervis et al., 1996; Sen and Stoffa, 1992a,b; Chunduru et al., 1995). Minister et al. (1995) find that evolutionary programming is useful for locating the hypocenter of an earthquake, especially when combined with simulated annealing.

Another inverse problem is determining the source of air pollutants, given what is known about monitored pollutants. Additional information includes the usual combination (percentages) of certain pollutants from different source regions and predominant wind patterns. The goal of the receptor inverse models is to target what regions, and even which sources, contribute the most pollution to a given receptor region. This process involves an optimization. Cartwright and Harris (1993) suggest that a genetic algorithm may be a significant advance over other types of optimization models for this problem when there are many sources and many receptors.

Evolutionary methods have also found their way into oceanographic experimental design. Barth (1992) showed that a genetic algorithm is faster than simulated annealing and more accurate than a problem-specific method for optimizing the design of an oceanographic experiment. Porto (1995) found that an evolutionary programming strategy was more robust than traditional methods for locating an array of sensors in the ocean after they have drifted from their initial deployment location.

Finally, Charbonneau (1995) gives three examples of uses of a genetic algorithm in astrophysics: modeling the rotation curves of galaxies, extracting pulsation periods of Doppler velocities in spectral lines, and optimizing a model of hydrodynamic wind.

7.2.1.5 Scheduling   The traveling salesman problem is a classic scheduling problem. Another similar problem found in industry is job shop scheduling. In this type of problem there are a certain number of jobs that must be done by a certain number of machines. Thus, several concurrent and possibly conflicting goals must be accomplished by a limited number of resources. Syswerda (1991) developed a genetic-algorithm-based scheduler for a Navy laboratory. He had to take into account resource constraints, time constraints, setup time, priority of tasks, and order of tasks. The traveling-salesman-type genetic algorithm found superior solutions compared to a random search given the same number of function evaluations for each algorithm.

A hybrid genetic algorithm consisting of a genetic algorithm that used a local minimum seeking algorithm in the final stages solved a hypothetical job shop scheduling problem 4% better than a genetic algorithm by itself (Uckun et al., 1993). The minimum seeking algorithm was invoked whenever the genetic algorithm found a best-ever cost. These results were averaged over 100 iterations but reported as a function of generation rather than number of cost function evaluations. Thus, the comparison between the hybrid genetic algorithm and simple genetic algorithm was not completely fair.

Ronald (1995) gives an excellent description of using genetic algorithms to solve many different scheduling problems. He provides explicit details on solving the traveling salesman, job shop scheduling, and open shop scheduling problems. The open shop scheduling problem is similar to the job shop scheduling problem, except the order in which the machines process a job is not important. He also describes the solution of other interesting scheduling problems using a genetic algorithm. These include very large-scale integration (VLSI) design, school bus routing, and resource routing.

7.2.2 Reference Materials

This book is an introduction to applying genetic algorithms to solve problems. We've assembled some introductory resources for the beginner.

The Internet.   Use a net browser such as Netscape. Do a search on "genetic algorithms." You will find more information than you care to find. The search may be narrowed by specifying additional search words. Some of the better web sites to start exploring include

```
http://www.aic.nrl.navy.mil
http://alife.santafe.edu
```

Also available from the internet is *The Hitch-Hiker's Guide to Evolutionary Computation* (Heitkotter and Beasley, 1994). This excellent up-to-date resource lists everything from where to get free software to many frequently asked questions (FAQ) about evolutionary programming.

Articles.   Many of the major scientific and engineering journals have published introductory articles on genetic algorithms. We listed several of them in the bibliographies at the end of the chapters. Holland's article in *Scientific American* is what got us started using genetic algorithms (Holland, 1992).

Books.   No, this is not the only genetic algorithm book. We've referenced most of the books currently available at the ends of the chapters. If you are interested in the theory of genetic algorithms, Holland (1975), Goldberg (1989), Mitchell (1996), Whitley (1993), and Schwefel (1995) will satisfy you. The other books [including Goldberg (1989)] are more application oriented.

Journals.   Several journals frequently publish articles on genetic algorithms. Some include:

Evolutionary Computation
IEEE Expert Intelligent Systems & Their Applications
IEEE Transactions on Systems, Man, and Cybernetics
BioSystems
Complex Systems
Machine Learning

Short Courses.   Several institutions offer short courses on genetic algorithms, including Georgia Institute of Technology and UCLA. In addition, many conferences offer short courses and tutorials on various aspects of genetic algorithms.

Software.   There is a lot of free software available. For the latest, consult Heitkotter and Beasley (1994). Some examples include:

MATLAB Genetic Algorithm Toolbox
Genetic Search Implementation System (GENESIS)
Genetic Algorithm for Numerical Optimization for Constrained Problems (GENOCOP)
Better to Use Genetic Systems (BUGS)
GENEsYs
Tool Kit for Genetics-Based Applications (TOLKIEN)

Conferences.   There are several conferences each year devoted to evolutionary programming, including

Conference on Genetic Programming
IEEE Conference on Evolutionary Computation
International Conference on Genetic Algorithms
Annual Conference on Evolutionary Programming
Foundations of Genetic Algorithms
Parallel Problem Solving from Nature

7.3 THE FUTURE

Our computer modeling skills lag the demand of our technical culture. Optimization saves money, time, and resources—the idea of doing more with less. Traditional optimization approaches have severe limitations, as shown in Chapter 1. A serious limitation of the traditional method is its serial nature. As computer architectures turn to parallel processing, the genetic algorithm will be there to become the optimization workhorse. It is the perfect algorithm for implementation on a parallel-processing computer. Simultaneously running several genetic algorithms and allowing migration between the populations of these algorithms can be done on parallel processors. In addition, the simple fact that the costs of individuals in a population can be simultaneously calculated tremendously speeds the convergence of a genetic algorithm.

Another interesting research area is the genetic algorithm with a subjective cost function. Can genetic algorithms be used to help in decision making? Perhaps someone interested in investing in the stock market could use a genetic algorithm to combine stock performance, stock types, and individual preferences to find an optimal portfolio for that individual. A genetic algorithm may also help the military develop strategies in war. The genetic algorithm could take into account supplies, reinforcements, civilian population, various attack/defense possibilities, and so forth. Maybe we could replace war with computer program competitions.

Many aspects of genetic algorithms still need investigation. Demystifying the optimal choice of genetic algorithm parameters would be an accomplishment. The studies outlined in Chapter 5 took toddler steps in that direction. Adaptive parameter settings may be the solution, but much work is needed in that arena. Hybridizing the genetic algorithm may provide a powerful alternative in many complex design problems. Steps toward proving the convergence of genetic algorithms would be another useful research area and could change current genetic algorithm implementations. Genetic programming is in its infancy. Perhaps not for long, though, since the idea of having a computer write computer code is extremely attractive.

Genetic algorithms are not a fad but are here to stay. Nature is not a bad model for our computing algorithms to follow. After all, the world has turned out pretty good. Maybe someday nature will find that global optimum.

BIBLIOGRAPHY

Altshuler, E. E., and D. S. Linden, 1997 Jan., "Design of a loaded monopole having hemispherical coverage using a genetic algorithm," *IEEE Trans. Antennas, Propagat. Syst.* **APS-45**(1), pp. 1–4.

Barth, N. H., 1992, "Oceanographic experiment design II: Genetic algorithms," *J. of Oceanic Atmos. Technol.* **9**, pp. 434–443.

Boag, et al., 1996 June, "Design of electrically loaded wire antennas using genetic algorithms," *IEEE Trans. Antennas Propagat. Syst.*, **APS-44**, pp. 687–695.

Carse, B., T. C. Fogarty, and A. Munro, 1995, "Evolutionary learning in computational ecologies: an application to adaptive distributed routing in communication networks," in T. C. Fogarty (Ed.), *Evolutionary Computing*, Berlin: Springer-Verlag, pp. 103–116.

Cartwright, H. M., and S. P. Harris, 1993, "Analysis of the distribution of airborne pollution using genetic algorithms," *Atmos. Environ.*, Part A, 27A, pp. 1783–1791.

Chambers, L., (Ed.), 1995, *Genetic Algorithms, Applications*, Vol. I, New York: CRC Press.

Charbonneau, P., 1995, "Genetic algorithms in astronomy and astrophysics," *Astrophys. J. Suppl. Ser.* **101**, pp. 309–334.

Chunduru, R. K., M. K. Sen, P. L. Stoffa, and R. Nagendra, 1995, "Non-linear inversion of resistivity profiling data for some regular geometrical bodies," *Geophy. Prospect.* **43**, pp. 979–1003.

Cox, L. A., Jr., L. Davis, and Y. Qiu, 1991, "Dynamic anticipatory routing in circuit-switched telecommunications networks," in L. Davis (Ed.), *Handbook of Genetic Algorithms*, New York: Van Nostrand Reinhold, pp. 124–143.

De Jong, K. A., 1975, "Analysis of the behavior of a class of genetic adaptive systems," Ph.D. Dissertation, The University of Michigan, Ann Arbor.

Fogel, L. J., A. J. Owens, and M. J. Walsh, 1966, *Artificial Intelligence Through Simulated Evolution*, New York: John Wiley & Sons.

Goldberg, D. E., 1989, *Genetic Algorithms in Search, Optimization, and Machine Learning*, Reading, MA: Addison-Wesley.

Harp, S. A., T. Samad, and A. Guha, 1991, "Towards the genetic synthesis of neural networks," in L. Davis (Ed.), *Handbook of Genetic Algorithms*, New York: Van Nostrand Reinhold, pp. 360–369.

Haupt, R. L., 1995 April, "An introduction to genetic algorithms for electromagnetics," *IEEE Antennas Propagat. Mag.* **37**(2), pp. 7–15.

Haupt, R. L., 1997 June, "Phase only adaptive nulling with genetic algorithms," *IEEE Trans. Antennas Propagat. Syst.* **APS-42**(7), pp. 918–924.

Haupt, R. L., 1994a July, "Thinned arrays using genetic algorithms," *IEEE Trans. Antennas Propagat. Syst.* **APS-42**(7), pp. 993–999.

Haupt, R. L., 1994b July, "Optimization of array antennas using genetic algorithms," *Proc. of the Progress in Electromagnetics Research Symposium*, Noordwijk, The Netherlands, p. 172.

Haykin, S., 1994, *Neural Networks a Comprehensive Foundation*, New York: Macmillan.

Heitkotter, J., and D. Beasley (eds.), 1994, *The Hitch-Hiker's Guide to Evolutionary Computation: A List of Frequently Asked Questions* (available via anonymous FTP from rtfm.mit..edu:/pub/usenet/news.answers/ai-faq/genetic).

Holland, J. H., 1975, *Adaptation in Natural and Artificial Systems*, Ann Arbor, MI: The University of Michigan Press.

Holland, J. H., 1992 July, "Genetic algorithms," *Sci. Am.*, pp. 66–72.

Jansen, D. J., and J. F. Frenzel, 1993 Oct. "Training product unit neural networks with genetic algorithms," *IEEE Expert Syst.*

Jervis, M., and P. L. Stoffa, 1993, "2-D migration velocity estimation using a genetic algorithm," *Geophys. Res. Lett.* **20**, pp. 1495–1498.

Jervis, M., M. K. Sen, and P. L. Stoffa, 1996, "Prestack migration veolocity estimation using nonlinear methods," *Geophysics* **60**, pp. 138–150.

Kapsalis, A., et al., 1995, "The radio link frequency assignment problem: A case study using genetic algorithms," in T. C. Fogarty, (Ed.), *Evolutionary Computing*, Berlin: Springer-Verlag, pp. 117–131.

Li, Q., et al., 1996, "Scattering center analysis of radar targets using fitting scheme and genetic algorithm," *IEEE Trans. Antennas Propagat. Syst.* **APS-42**(2), pp. 198–207.

Linden, D. S., and E. E. Altshuler, 1996, "The design of yagi antennas using a genetic algorithm," USNC/URSI Radio Science Meeting, Baltimore, MD, July 21–26, p. 283.

Michielssen, E., et al., 1993 June/July, "Design of lightweight, broad-band microwave absorbers using genetic algorithms," *IEEE Trans. Microwave Theory Tech.* **MTT-41**, pp. 1024–1031.

Miller, G. F., P. M. Todd, and S. U. Hegde, 1989, "Designing neural networks using genetic algorithms," *Proc. of the Third International Converence on Genetic Algorithms*, pp. 379–384.

Minister, J-B. H., N. P. Williams, T. G. Masters, J. F. Gilbert, and J. S. Haase, 1995, "Application of evolutionary programming to earthquake hypocenter determination," in *Evolutionary Programming: Proc. of the Fourth Annual Conference on Evolutionary Programming*, pp. 3–17.

Mitchell, M., 1996, *An Introduction to Genetic Algorithms*, Cambridge, MA: The MIT Press.

Ng, S. C., et al., 1996, "The Genetic Search Approach," *IEEE Signal Process. Mag.* **13**, pp. 38–46.

Porto, V. W., 1995, "Non-acoustic sensor array localization using evolutionary programming," in *Evolutionary Programming: Proc. of the Fourth Annual Conference on Evolutionary Programming*, pp. 19–32.

Porto, V. W., D. B. Fogel, and L. J. Fogel, 1995, June, "Alternative neural network training methods," *IEEE Expert Syst.*

Rechenberg, I., 1965 August, "Cybernetic solution path of an experimental problem," *Royal Aircraft Establishment, Library Translation 1122*, Farnborough, Hants, England.

Ronald, S., 1995, "Routing and scheduling problems," in L. Chambers, (Ed.), *Genetic Algorithms, Applications* Vol. I, New York: CRC Press, pp. 367–430.

Schwefel, H., 1995, *Evolution and Optimum Seeking*, New York: John Wiley.

Sen, M. K., and P. L. Stoffa, 1992a, "Rapid sampling of model space using genetic algorithms: examples from seismic waveform inversion," *Geophys. J. Int.* **108**, pp. 281-292.

Sen, M. K., and P. L. Stoffa, 1992b, "Genetic inversion of AVO," *Geophy.: Leading Edge Explor.*, pp. 27–29.

Sen, M. K., and P. L. Stoffa, 1996, "Bayseian inference, Gibbs' sampler and uncertainty estimation in geophysical inversion," *Geophys. Prospect.* **44**, pp. 313–350.

Syswerda, G., 1991, "Schedule optimization using genetic algorithms," in L. Davis (Ed.), *Handbook of Genetic Algorithms*, New York: Van Nostrand Reinhold, pp. 332–349.

Tang, K. S., et al., 1996, "Genetic algorithms and their applications," *IEEE Signal Process. Mag.* **13**, pp. 22–37.

Tang, K. S., et al., 1995, "GA approach to multiple objective optimization for active noise control," *Algorithms and Architectures for Real-Time Control 95*, Belgium, 31 May–2 June, pp. 13–19.

Uckun, S., et al., 1993 Oct., "Managing genetic search in job shop scheduling," *IEEE Trans. Expert Syst.*, pp. 15–24.

Whitley, L. D., (ed.), 1993, *Foundations of Genetic Algorithms 2*, San Mateo, CA: Morgan Kaufmann.

APPENDIX A

PSEUDOCODES

We don't provide computer code in this book. One author is a MATLAB convert, while the other is a die-hard Fortran programmer. Between us we know a smattering of C, Pascal, Algol, PL1, Mathematica, Maple, and MathCad. Rather than risk divorce over which code to provide, we resorted to pseudocode. Hopefully enough detail is provided to let the reader easily write a customized genetic algorithm. Some guidelines for reading the pseudocode:

- A matrix is a variable with all capital letters
- A vector is a variable with the first letter capitalized
- A scalar is all lowercase letters
- A function provided by the user is in boldface
- A function described by pseudocode is italicized
- % indicates a program comment

User-provided subroutines are described below:

- random(r, c)—generates an $r \times c$ matrix of uniformly distributed random numbers.
- round(*)—rounds numbers to nearest integer.

- costfunction(CHROMOSOMES)—returns a column vector with the cost associated with each chromosome or row in the matrix CHROMOSOMES.
- sort(Cost, CHROMOSOMES)—sorts the Cost vector and associated chromosomes from lowest cost in row 1 to highest cost in the last row. It also truncates the vector and matrix to the first popsize rows.
- min—finds the minimum value of the vector.
- mean—finds the average value of the vector.
- std—finds the standard deviation of the elements in the vector.
- roundup—rounds numbers to next highest integer.

Most math packages provide some form of the above functions. Canned subroutines for programming languages are generally available for most of these functions as well.

The main genetic algorithm and pairing function are the same for the binary and continuous versions. However, the mating and mutation functions are quite different, so two versions are provided.

Pseudocode for a Binary Genetic Algorithm

```
%_____
% Define variables
%_____
maxiterations=?          % maximum number of iterations
ipopsize=?               % population size of generation 0
popsize=?×ipopsize       % population size of generations 1 through...
keep=?×popsize           % number of chromosomes kept for mating
bits=?                   % total number of bits in a chromosome
mutaterate=?             % mutation rate
-------------------------------------------
%
% Create the initial population, evaluate costs, and sort
%_____
CHROMOSOMES=round(random(ipopsize,bits))
                         % matrix of random 1s & 0s

----------------------
%
% Let the generations begin!
% Cost - vector containing the costs
% sort - sorts & truncates costs & chromosomes
```

```
%------------------
gen#=0              % initial generation
quit='no'           % convergence check
while gen#<maxiterations & quit='no'
gen#=gen#+1         % increment the generation number
Cost=costfunction(CHROMOSOMES)
[Cost,CHROMOSOMES]=sort(Cost,CHROMOSOMES)

%------------------
%
% Evaluate cost statistics
%----------------
Mincost(gen#)=min(Cost)       % minimum cost
Meancost(gen#)=mean(Cost)     % mean cost
Stndcost(gen#)=std(Cost)      % standard deviation of cost

%---------------------------------------------
%
% The chromosomes are paired and offspring are produced
%---------------------------------------------
[Mom Dad]=pair(CHROMOSOMES,Cost,keep,popsize)
CHROMOSOMES=matebin(Mom,Dad,CHROMOSOMES,keep,
                         popsize,bits)

%----------------------
%
% mutate the population with (popsize × N × mutaterate) mutations
%--------------------
CHROMOSOMES=mutatebin(CHROMOSOMES,mutaterate,
                         popsize,bits)

%-------------------
%
% Check for convergence
%----------------
if Mincost(gen#)<? and/or Meancost(gen#)<?
               and/or Stndcost(gen#)<?⇒quit='yes'
end
```

Pseudocode for Pairing

```
function [Mom,Dad]=pair(CHROMOSOMES,Cost,keep,popsize)
%----------------
% Selects one of three options for calculating the probability
%----------------
```

replacements=(popsize-keep)/2   % # CHROMOSOMES needing
                                  % replaced

$$
\text{Prob}(n) = \begin{cases}
\dfrac{n}{\sum_{r=1}^{\text{replacements}} r} \\[2ex]
\dfrac{|\text{cost}(n)|}{\sum_{r=1}^{\text{replacements}} |\text{cost}(r)|} \\[2ex]
\dfrac{\text{cost}(n) - \min\{\text{cost}(\text{replacements} + 1)\}}{\sum_{r=1}^{\text{replacements}} |\text{cost}(r) - \min\{\text{cost}(\text{replacements} + 1)\}|}
\end{cases}
$$

% cummulative probabilities
Odds=[0, Prob(1), Prob(1)+Prob(2), ..., $\sum_{n=1}^{\text{replacements}}$Prob(n)]

%
% Roll dice for parents
%---------------
Pick1=**random**(1,replacements)   % vector of random numbers for Mom
Pick2=**random**(1,replacements)   % vector of random numbers for Dad

----
%
% finds the two mates
%----------------------
ic=1   % initialize counter
while ic<replacements   % counter must be less than
                                    % replacement number

%
% when one of the random Picks falls inside a cummulative probability
% bin, the chromosome associated with that bin is selected as a parent
%--------------------------
for id=2:keep+1
if Pick1(ic)<Odds(id) & Pick1(ic)>Odds(id−1) → Mom(ic)=id−1
if Pick2(ic)<Odds(id) & Pick2(ic)>Odds(id−1) → Dad(ic)=id−1
end
ic=ic+1   % increment counter
end

Pseudocode for Binary Mating

function CHROMOSOMES=*matebin*(Mom,Dad,CHROMOSOMES,
                                    keep,popsize,bits)

```
----------------------------
%
% selects a crossover point
% roundup rounds to next highest integer
%_____
Xpoint=roundup{(N−1)×random(1,M)}

-------------------------------------------
%
% row indx contains first offspring
% row indx+1 contains second offspring
% mom - vector containing row numbers of first parent
% dad - vector containing row numbers of second parent
%_____
for ic=1:popsize
indx=2×(ic-1)+1
CHROMOSOMES(keep+indx,1→popsize)=
    [CHROMOSOMES(Mom(ic),1→Xpoint(ic)),
    CHROMOSOMES(Dad(ic),Xpoint(ic)+1 →popsize)]
CHROMOSOMES(keep+indx+1,1→popsize)=
    [CHROMOSOMES(Dad(ic),1→Xpoint(ic)),
    CHROMOSOMES(Mom(ic),Xpoint(ic)+1 →popsize)]
end
```

Pseudocode for Binary Mutation

```
function CHROMOSOMES=mutatebin(CHROMOSOMES,mutaterate,
                                popsize,bits)

-------------------------------------------
%
% Inside a loop iterating over the number of mutations, a random
% bit in the population is selected and changed from a 1 to a 0 or
% from a 0 to a 1.
%_____
#mutations=roundup(popsize×bits×mutaterate)
                                    % number of mutations
for ic=1→#mutations
row=roundup((popsize-2)×random(1,1))+1    % random row
col=roundup((bits-2)×random(1,1))+1       % random column
CHROMOSOMES(row,col)=CHROMOSOMES(row,col)−1
                                    % mutation
end
```

Pseudocode for a Continuous Parameter Genetic Algorithm

```
%_____
% Define variables
%_____
popsize=?×ipopsize      % population size of generations 1 through...
maxiterations=?         % maximum number of iterations
ipopsize=?              % population size of generation 0
popsize=?×ipopsize      % population size of generations 1 through...
keep=?×popsize          % number of chromosomes kept for mating
pars=?                  % total number of parameters in a chromosome
mutaterate=?            % mutation rate
hi=?                    % maximum parameter value
lo=?                    % minimum parameter value

_____
%
% Create the initial population, evaluate costs, and sort
%_____
CHROMOSOMES=(hi−lo)×random(ipopsize,bits)+lo
                        % matrix of random 1s & 0s

_____
%
% Let the generations begin!
%_____
gen#=0                  % initial generation
quit='no'               % convergence check
while gen#<maxiterations & quit='no'
gen#=gen#+1             % increment the generation number
Cost=costfunction(CHROMOSOMES)
[Cost,CHROMOSOMES]=sort(Cost,CHROMOSOMES)

_____
%
% Evaluate cost statistics
%_____
Mincost(gen#)=min(Cost)      % minimum cost
Meancost(gen#)=mean(Cost)    % mean cost
Stndcost(gen#)=std(Cost)     % standard deviation of cost

_____
%
% The chromosomes are paired and offspring are produced
%_____
```

[Mom Dad]=*pair*(CHROMOSOMES,Cost,keep,popsize)
CHROMOSOMES=*matecon*(Mom,Dad,CHROMOSOMES,keep,
popsize,pars)

%
% mutate the population with (popsize×N×mutaterate) mutations
%_____
CHROMOSOMES=*mutatecon*(CHROMOSOMES,mutaterate,popsize,
pars,hi,lo)

%
% Check for convergence
%_____
if Mincost(gen#)<? and/or Meancost(gen#)<? and/or Stndcost(gen#)
<?⇒quit='yes'
end

Pseudocode for Continuous Parameter Mating

function CHROMOSOMES=*matecon*(Mom,Dad,CHROMOSOMES,
keep,popsize,pars)

%
% row indx contains first offspring
% row indx+1 contains second offspring
% mom - vector containing row numbers of first parent
% dad - vector containing row numbers of second parent
%_____
for ic=1→popsize
indx=2×(ic−1)+1
alpha=**roundup**{**random**×pars}
beta=**random**
CHROMOSOMES(keep+indx,alpha)
= CHROMOSOMES(Mom(ic),alpha)
− beta ×[CHROMOSOMES(Mom(ic),alpha)
− CHROMOSOMES(Dad(ic),alpha)]
CHROMOSOMES(keep+indx+1,alpha)
= CHROMOSOMES(Dad(ic),alpha)
+ beta ×[CHROMOSOMES(Mom(ic),alpha)
− beta × CHROMOSOMES(Dad(ic),alpha)]
if alpha> 1 & alpha<pars

CHROMOSOMES(keep+indx,alpha+1→pars)
    = CHROMOSOMES(keep+indx+1,alpha+1→pars)
CHROMOSOMES(keep+indx+1,alpha+1→pars)
    = CHROMOSOMES(keep+indx,alpha+1→pars)
end
end

Pseudocode for Continuous Parameter Mutation

function CHROMOSOMES=*mutatecon*(CHROMOSOMES,maxval,
                                    mutaterate,popsize,pars,hi,lo)

%
% Inside a loop iterating over the number of mutations, a random
% parameter in the population is selected and replaced by a new
% random parameter.
%---
#mutations=**roundup**(popsize×pars×mutaterate)   % # mutations

for ic=1→#mutations
row=**roundup**(popsize×random)+1   % random row
col=**roundup**(pars×random)   % random column
CHROMOSOMES(row,col)= (hi−lo)×**random**+**lo**   % mutation end

GLOSSARY

This book has a mixture of biology, mathematics, and computer terms. This glossary is provided to help the reader keep the terminology straight.

Allele: The value of a gene. In biology, one of the functional forms of a gene.

Age of a Chromosome: The number of generations that a chromosome has existed.

Building Block: A group of genes that give a chromosome a high fitness.

Chromosome: An array of parameters or genes that is passed to the cost function.

Comma Strategy: The process in which the parents are discarded and the offspring compete.

Converge: Arrive at the solution. A gene is said to have converged when 95% of the chromosomes contain the same allele for that gene. GAs are considered converged when they stop finding better solutions.

Convergence Rate: The speed at which the algorithm approaches a solution.

Cooperation: The behavior of two or more individuals acting to increase the gains of all participating individuals.

Cost: Output of the cost function.

Cost Function: Function to be optimized.

Cost Surface: Hypersurface that displays the cost for all possible parameter values.

Crowding Factor Model: An algorithm in which an offspring replaces a chromosome that closely resembles the offspring.

Crossover: An operator that forms a new chromosome from two parent chromosomes by combining part of the information from each.

Crossover Rate: A number between zero and one that indicates how frequently crossover is applied to a given population.

Cycle Crossover: A method of crossover for permutation problems in which a point is initially chosen for exchange of genes between the parents; then the remainder of the operator involves "cycling" through the parents to eliminate doubles in the offspring.

Darwinism: Theory founded by Charles Darwin that evolution occurs through random variation of heritable characteristics, coupled with natural selection (survival of the fittest).

Deceptive Functions: Functions that are difficult for the genetic algorithm to optimize.

Diploid: A pair of chromosomes carrying the full genetic code of an organism.

Elitism: The chromosome with the best cost is kept from generation to generation.

Environment: That which surrounds an organism.

EP: See evolutionary programming.

Epistasis: The interaction or coupling between different parameters of a cost function. The extent to which the contribution to fitness of one gene depends on the values of other genes. Highly epistatic problems are difficult to solve, even for GAs. High epistasis means that building blocks cannot form, and there will be deception.

ES: See evolution strategy.

Evolution: A series of genetic changes in which living organisms acquire the characteristics that distinguish it from other organisms.

Evolution Strategy (ES): A type of evolutionary algorithm developed in the early 1960s in Germany. It employs continuous parameters. In its original form, it relied on mutation as the search operator, and a population size of one. Since then it has evolved to share many features with genetic algorithms.

Evolutionary Computation: Encompasses methods of simulating evolution on a computer. The term is relatively new and represents an effort to bring together researchers who have been working in closely related fields but following different paradigms. The field is now seen as including research in genetic algorithms, evolution strategies, evolution programming, and artificial life.

Evolutionary Programming (EP): An evolutionary algorithm developed in the mid-1960s. It is a stochastic optimization strategy, which is similar to genetic algorithms, but dispenses with both "genomic" representations and with crossover as a reproduction operator.

Exploitation: Information from visited points in the search space determines which points might be profitable to visit next.

Exploration: Randomly exploring new regions of a search space. Helps algorithm stay out of local minima. Problems that have many local maxima can sometimes only be solved by this sort of random search.

Extremum: A maximum or a minimum.

Fitness: Opposite of cost. A value associated with a chromosome that assigns a relative merit to that chromosome.

Fitness Function: Has the negative output of the cost function. Mathematical subroutine that assigns a value or fitness to a set of parameters.

Fitness Landscape: The inverted cost surface. The hypersurface obtained by applying the fitness function to every point in the search space.

Function Optimization: Process of finding the best extremum of a function.

GA: See genetic algorithm.

Gamete: Cells with haploid chromosomes that carry genetic information from their parents for the purposes of sexual reproduction. In animals, male gametes are called sperm and female gametes are called ova.

Gene: The binary encoding of a single parameter. A unit of heredity that is transmitted in a chromosome and controls the development of a trait.

Gene Flow: Introduction of new genetic information by introducing new individuals into the breeding population.

Gene Frequency: The incidence of a particular allele in a population.

Generation: One iteration of the genetic algorithm.

Generation Gap: A generation gap algorithm picks a subset of the current population for mating.

Genetic Algorithm (GA): A type of evolutionary computation devised by John Holland. It models the biological genetic process by including crossover and mutation operators.

Genetic Drift: Changes in gene/allele frequencies in a population over many generations, resulting from chance rather than selection. Occurs most rapidly in small populations. Can lead to some genes becoming "extinct," thus reducing the genetic variability in the population.

Genetic Programming: Genetic algorithms applied to computer programs.

Genotype: The genetic composition of an organism. The information contained in the genome.

Genome: The entire collection of genes (and hence chromosomes) possessed by an organism.

Global Minimum: True minimum of the entire search space.

Global Optimization: Finding the true optimum in the entire search space.

Gray Code: Binary representation of a parameter in which only one bit changes between adjacent quantization levels.

Hamming Distance: The number of bits by which two codes (chromosomes) differ.

Haploid Chromosome: A chromosome consisting of a single sequence of genes. The number of chromosomes contained in the gamete. Half the diploid number.

Hard Selection: When only the best available individuals are retained for generating future progeny.

Heterozygous: The members of a gene pair are different.

Hillclimbing: Investigates adjacent points in the search space, and moves in the direction giving the greatest increase in fitness. Exploitation techniques that are good at finding local extrema.

Homozygous: Both members of a gene pair are the same.

Hybrid Genetic Algorithm: A genetic algorithm combined with other optimization techniques.

Individual: A single member of a population that consists of a chromosome and its cost function.

Inversion: A reordering operator that works by selecting two cut points in a chromosome, and reversing the order of all the genes between those two points.

Kinetochore: The random point on a chromosome at which crossover occurs.

Lamarckism: Theory of evolution that preceded Darwin's. Lamarck believed that evolution resulted from the inheritance of acquired characteristics. The skills or physical features acquired by an individual during its lifetime can be passed on to its offspring.

Lifetime: How many generations a chromosome stays in the population until it is eliminated.

Local Minimum: A minimum in a subspace of the search space.

Mating Pool: A subset of the population selected for potential parents.

Meiosis: The type of cell division that occurs in sexual reproduction.

Messy Genetic Algorithm: Invented to conquer deceptive functions. The first step (primordial phase) initializes the population so it contains all possible building blocks of a given length. In this phase only reproduction is used to enrich the set of good building blocks. The second step (juxtaposition phase) uses various genetic operators to converge.

Migration: The transfer of the genes of an individual from one subpopulation to another.

Mitosis: Reproduction of a single cell by splitting. Asexual reproduction.

Multimodal: Cost surface with multiple minima.

Mutation: A reproduction operator that randomly alters the values of genes in a parent chromosome.

Mutation Rate: Percentage of bits in a population mutated each iteration of the GA.

Natural Selection: The most fit individuals reproduce, passing their genetic information on to their offspring.

Nelder Mead Downhill Simplex Algorithm: A nonderivative, robust local optimization method developed in 1965.

Niche: The survival strategy of an organism (grazing, hunting, on the ground, in trees, etc.). Species in different niches (e.g., one eating plants, the other eating insects) may coexist side-by-side without competition. If two species occupying the same niche are brought together, there will be competition, and eventually the weaker of the two species will be made extinct. Hence, diversity of species depends on them occupying a diversity of niches or on geographical separation. In evolutionary computations, we often want to maintain diversity in the population. Sometimes a fitness function may be known to be multimodal, and we want to locate all the peaks. We may consider each peak in the fitness function as analogous to a niche.

Object Variables: Parameters that are directly involved in assessing the relative worth of an individual.

Off-line Performance: An average of all costs up to the present generation. It penalizes the algorithm for too many poor costs, and rewards the algorithm for quickly finding where the lowest costs lie.

Offspring: An individual generated by any process of reproduction.

On-line Performance: The best cost found up to the present generation.

Optimization: The process of iteratively improving the solution to a problem with respect to a specified objective function.

Order-based Problem: A problem where the solution must be specified in terms of an arrangement (e.g. a linear ordering) of specific items, for example, traveling salesman problem, computer process scheduling. Order-based problems are a class combinatorial optimization problems in which the entities to be combined are already determined.

Order Crossover (OX): A crossover method for dealing with a permutation operator that strives to preserve portions of the absolute ordering of each parent.

Ontogeny: The developmental history (birth to death) of a single organism.

Parallel Genetic Algorithm: A genetic algorithm written to run on a parallel-processing computer.

Parent: An individual that reproduces to generate one or more other individuals, known as offspring, or children.

Partially Matched Crossover (PMX): A reordering operator where two crossover points are chosen, the values between these points exchanged, then a careful procedure followed to eliminate any repeated numbers from the solution.

Performance: Usually some statistical evaluation of the cost or fitness over all generations.

Permutation Problem: A problem that involves reordering a list.

Phenotype: The environmentally and genetically determined traits of an organism. That trait actually observed.

Phylogeny: The developmental history of a group of organisms.

Plus Strategy: The process in which the parents and offspring compete.

Point Mutation: Alters a single feature to some random value.

Population: A group of individuals that interact (breed) together.

Quadratic Surface: Bowl-shaped surface.

Random Seed: A number passed to a random number generator that the random number generator uses to initialize its production of random numbers.

Recombination: Combining the information from two parent chromosomes via **crossover**.

Reordering: Changing the order of genes in a chromosome to try to bring related genes closer together and aid in the formation of building blocks.

Reproduction: The creation of offspring from two parents (sexual reproduction) or from a single parent (asexual reproduction).

Reproduction Operator: The algorithmic technique used to implement reproduction.

Roulette Wheel Selection: Picks a particular population member to be a parent with a probability equal to its fitness divided by the total fitness of the population.

Scaling: Used to bring the range of costs into a desirable range. Often used to make all the costs positive or negative.

Schema (pl. schemata)**:** Bit pattern in a chromosome. For instance, the patterns 1100110 and 1000011 both match the schema 1##0011, where # indicates a 1 or a 0.

Schema Theorem: A GA gives exponentially increasing reproductive trials to schemata with above average fitness. Because each chromosome contains a great many schemata, the rate of schema processing in the population is very high, leading to a phenomenon known as implicit parallelism. This gives a GA with a population of size N a speedup by a factor of N cubed, compared to a random search.

Search Space: All possible values of all parameters under consideration.

Search Operators: Processes used to generate new chromosomes.

Seeding: Placing good guesses to the optimum parameter values in the initial population.

Selection: The process of choosing parents for reproduction (usually based on fitness).

Self-Adaptation: The inclusion of a mechanism not only to evolve the object variables of a solution, but simultaneously to evolve information on how each solution will generate new offspring.

Single-Point Crossover: A random point in two chromosomes (parents) is selected. Offspring are formed by joining the genetic material to the right of the crossover point of one parent with the genetic material to the left of the crossover point of the other parent.

Simulation: The act of modeling a process.

Soft Selection: Some inferior individuals in a population are allowed to survive into future generations.

Speciation: The process of developing a new species. The most common form of speciation occurs when a species is geographically separated from the main population long enough for their genes to diverge due to differences in selection pressures or genetic drift. Eventually, the genetic differences are great enough for the subpopulation to become a new species.

Species: A group of organisms that interbreed and are reproductively isolated from all other groups. A subset of the population.

Steady-State Genetic Algorithm: Every new offspring produced replaces a high cost chromosome in the population.

Subpopulation: A subset of the main population in which individuals may only mate with others in the same subset.

Survival of the Fittest: Only the individuals with the highest fitness value survive.

Tournament Selection: Picks a subset of the population at random, then selects the member with the best fitness.

Two-point Crossover: A crossover scheme that selects two points in the chromosome and exchanges bits between those two points.

Uniform Crossover Randomly assigns the bit from one parent to one offspring and the bit from the other parent to the other offspring.

INDEX